ANGST AND THE ABYSS

American Academy of Religion Academy Series

edited by
Carl A. Raschke

Number 49

ANGST AND THE ABYSS
The Hermeneutics of Nothingness
by
David K. Coe

David K. Coe

ANGST AND THE ABYSS
The Hermeneutics of Nothingness

Scholars Press
Chico, California

ANGST AND THE ABYSS
The Hermeneutics of Nothingness

by
David K. Coe
Ph.D., 1981, University of Hawaii
Manoa, Hawaii

Library of Congress Cataloging in Publication Data

Coe, David K., 1941–
Angst and the abyss.

(Academy Series / American Academy of Religion ; no. 49)
Bibliography: p.
1. Anxiety—History. 2. Philosophical anthropology—History. 3. Ontology—History. I. Title. II. Series: American Academy of Religion academy series ; no. 49.
BD450.C59 1985 128'.3 85-8129
ISBN 0-89130-862-8 (alk. paper)
ISBN 0-89130-863-6 (pbk. : alk. paper)

Printed in the United States of America
on acid-free paper

TO
BARBARA ANN COE

CONTENTS

ACKNOWLEDGEMENTS

I wish to express deep gratitude to several persons who made this work possible. To Professor J. L. Mehta, whose friendship and kindness helped open many long-hidden doors to the true meaning of Heidegger's path of thinking, my deepest thanks.

To Maxine Kurtz, who read the early drafts and offered useful criticisms, I owe much gratitude. To Dr. Michael S. March, who demanded that the dream of this work be re-awakened to reality, I am profoundly grateful. To Tom Abrums, who in 1960 attempted to show me the path to writing more than subjectivist scribbles, my most sincere thanks. Finally, to Larry Stallings, my most profound thanks: this effort was, after all, a result of his deep-seated quest for the meaning of Angst.

INTRODUCTION

The general purpose of this essay is to inquire into the role of Angst in the philosophy of the West. Here, in this "Introduction," we hope to suggest a kind of cognitive map for what follows in the body of this work.

At the outset, then, we must state unequivocally that this essay is neither a work in depth-psychology nor in what has become known as "existentialism." Indeed, we are decidedly not interested in discussing "anxiety" per se as either of these two disciplines use that term to describe a psychological state or a general feeling of "uneasiness" that creeps in like some mute fog to drift through the darker moments of human existence. On the contrary, to allow primordial Angst to be revealed in its own right requires us to get behind and beneath both the depth psychological and the "existentialist" concepts of "anxiety" to the rock-bottom ontological condition grounding these concepts. So to illuminate the horizons of this work to an ontological inquiry we designate the underlying condition of the possibility of everyday or psychological anxiety by the key term, "primordial Angst."

The means by which we hope to secure a revelation of primordial Angst is grounded in Martin Heidegger's magnum opus of 1927, *Sein and Zeit*. Heidegger's "method," which he called "hermeneutic phenomenology" or the phenomenology of understanding and interpretation with regard to human beings, is the basis of our own "methodology," one further refined by Hans-Georg Gadamer's *Wahrheit and Methode*./1/ But our own specific contribution to this methodology, and this is a point the reader might want to keep in mind, is to distinguish between two spheres of the enterprise: the "macro-hermeneutic" and the "micro-hermeneutic."

In the macro-hermeneutic approach, the approach of our first and final chapters, we attempt grasping primordial Angst in the broadest context of Western thought. More specifically, in

Chapter I, we sketch out how primordial Angst was revealed from pre-philosophical times up through the late medieval and early Renaissance period of Western civilization.

The next three chapters, however, are markedly different in scope and intent; they are *micro-hermeneutical* chapters. By "micro-hermeneutical" we mean our focus goes from the broadest possible perspective on Angst to one almost microscopic in detail, one exploring the precise horizons of Angst's disclosure within the purview of four individual philosophers: Kierkegaard, Heidegger, Sartre, and Tillich. Thus in these chapters we are concerned with topics such as the various philosophical influences on these thinkers so that Angst may be revealed within the proper context of their respective philosophical positions.

The purpose of this detailed, almost microscopic analysis is to ground the macro-hermeneutical interpretation of primordial Angst offered in our final chapter; for without the "micro-hermeneutics" of Chapters II, III, and IV, we would have no sound foundation upon which an adequate "macro-hermeneutical" interpretation of primordial Angst could be offered.

In this introduction, we should clear up a preliminary question: Why have we not translated Angst as either "anxiety," "dread," or "anguish," the three most popular renderings of Angst?

Angst has been left untranslated because there *is* not, nor *can* there be, an adequate rendering which fully conveys Angst's subtleties. Hence, we submit that rather than a translation, Angst requires a phenomenological description and interpretation. But why do the translations of Angst offered previously strike us as being inadequate? Let us examine them hermeneutically.

First, regarding "anxiety," the most frequent rendering of "Angst," John MacQuarrie and Edward Robinson, the translators of *Sein und Zeit*, admit their choice of this translation was really the least objectionable of several alternatives. Others they considered were "dread," "uneasiness," and "malaise." They admit, however, that "in some ways 'uneasiness' or 'malaise' would be more appropriate still" for what Heidegger had in mind by Angst./2/ In a later work, MacQuarrie confirms this view:

> It must be acknowledged . . . that it is not altogether a satisfactory translation. It suggests too much day-to-day anxieties of ordinary life rather than the rare and subtle emotions that the existentialists wish to designate by the German word *Angst*./3/

Earlier Macquarrie and Robinson justified "anxiety" for "Angst" on the grounds that Angst has generally been translated as "anxiety" in post-Freudian psychological literature./4/ What they are admitting, in our view, is an *appeal to convention,* a circumstance that has profound, far-reaching, and regrettable consequences. Not only does this appeal violate one of the basic cannons of hermeneutics articulated by Heidegger, but it radically misleads the readers of *Being and Time* into interpreting Angst simply as *psychological anxiety* and nothing more. This casual linking up of Heidegger's term "Angst" with post-Freudian psychological literature, then, colors Angst's meaning, robbing Angst at the outset of its fully ontological richness.

Moreover, from the standpoint of hermeneutical phenomenology itself, the "anxiety" interpretation of Angst is both trivial and misleading. Heidegger has shown that Angst is part of the universal condition of being human; an ontological "category," or what he calls an "*Existenzial*" of that being whose Being is an issue for it, namely Dasein. Very briefly, these ontological categories of Dasein include understanding (*Verstehen*), disposition or attunement (*Befindlichkeit*), and speech or discourse (*Rede*). Each is present in Dasein and each characterizes Dasein's essential character: Dasein's essence lies in its existence./5/ Heidegger insists that these are true *ontological* "categories" of Dasein rather than Dasein's ontical characteristics. As MacQuarrie and Robinson note, "Ontological inquiry is concerned primarily with Being, ontical inquiry is concerned with *entities* and the facts about them."/6/

From Heidegger's perspective then, the use of "anxiety " for Angst places Angst directly under, indeed squarely within, the purview of ontic psychological experience rather than the ontological and a priori conditions which make possible psychological experience. But Angst is precisely ontological. It is Dasein's primordial pre-disposition or attunement to Being (*Grundbefindlichkeit*). Thus, to see Angst as mere "anxiety" is to confuse the ontological condition of the possibility of feeling in general with a specific and derived mode of feeling Heidegger calls "uncanniness."

The second most popular translation of Angst is "dread." Using "dread" for Angst gives rise to pressing difficulties even beyond the use of "anxiety," as we shall see. Walter Lowrie,/7/ translator of Søren Kierkegaard's work on Angst, *Begrebnet Angest*, informs us that some mysterious figure named Professor Hollander was responsible for first translating the Danish word *Angest* (or Angst in contemporary Danish) as "dread." In 1924 Hollander was responsible for publishing the first English translations of some Kierkegaard fragments./8/ Lowrie goes on to point out that after a desperate search for a better translation, everyone simply agreed to use Hollander's "dread" for *Angest.*/9/ Nonetheless, Lowrie himself confesses: "We have no word which adequately translates Angst."/10/

Six years later, Lowrie compounded this mistranslation error by allowing Rollo May to use "anxiety" in place of "dread." May reports:

> The question is whether the psychological meaning of "anxiety" . . . is not very close—in fact much closer than the term "dread"—to what Kierkegaard meant by Angst. Professor Paul Tillich, who was familiar with both the psychological meaning of Angst and Kierkegaard's works, believed this to be true./11/

To Paul Tillich's interpretation of Angst we shall return in Chapter IV, but here we must point out that May appeals to the authority of Tillich, another form of appeal to convention, an appeal grounded in a popular psychological conception of Angst or even a fanciful idea of its primary meaning.

The third most popular translation of Angst comes from the French *angoisse*, meaning "anguish." The popularity of translating Angst as "anguish" stems from Jean Paul Sartre's second order interpretation of Heidegger's Angst, an interpretation provided in *L'Etre et le neant*,/12/ translated in 1946 by Hazel Barnes as *Being and Nothingness.*/13/ *Angoisse* is indeed etymologically related to Angst as is the English word "anguish." Etymology notwithstanding, *Webster's Third New International Dictionary* defines "anguish" as: "extreme pain either of body or mind: excruciating distress."/14/ This reference to great pain is, as far as we know, neither present in Angst as interpreted by Kierkegaard or Heidegger, nor, for that matter, in Sartre's own work. Thus, while Barnes was etymologically correct in translating *angoisse* as "anguish," her translation is somewhat misleading.

From this brief and extremely provisional analysis we believe the three popular translations of Angst leave much to be desired. Not one is either appropriate or adequate, in our view, to convey the subtle complexity of the Angst phenomenon. Thus we propose to show how Angst requires an ontological and phenomenological description and interpretation to demonstrate its meaning in Western philosophical thought. Such is the task of the present essay: to understand and interpret Angst and the abyss metaphor from both a macro-hermeneutical and a micro-hermenutical perspective. But the guiding thought governing both our micro and macro hermeneutics is expressed by Heidegger as follows ". . . our first, last and constant task is never to let [our hermeneutics] *be determined by fancies or popular conceptions.*" /15/

NOTES

/1/ The method we intend to employ in this essay, namely the hermeneutic phenomenological method, was first discussed by Martin Heidegger in his work *Sein und Zeit, Erste Halfte* in the *Jahrbuch fur Philosophie und phaenomenologische Forschung*, 8 (Halle: Niemeyer Verlag, 1927). The method has been considerably expanded upon by Hans-Georg Gadamer in his work *Wahrheit und Methode* (Tübingen: Paul Siebeck, 1960).

/2/ Martin Heidegger, *Being and Time*, trans. by John Macquarrie and Edward Robinson (New York: Harper & Row, 1962), p. 227, note.

/3/ John MacQuarrie, *Existentialism* (Harmondsworth, Middlesex: Penguin Books, 1973), pp. 164–65.

/4/ Heidegger, *Being and Time*, p. 227, note.

/5/ Ibid., p. 67.

/6/ Ibid., p. 31, note 3.

/7/ Søren Kierkegaard, *The Concept of Dread*, trans. by Walter Lowrie (Princeton: Princeton University Press, 1944).

/8/ Ibid., "Translator's Preface," p. X.

/9/ Ibid.

/10/ Ibid.

/11/ Rollo May, *The Meaning of Anxiety*, revised edition (New York: W. W. Norton and Company, 1977), p. 37, note 39.

/12/ Jean Paul Sartre, *L'Etre el le neant: Essai d'ontologic phenomenologique* (Paris: Gallimard, 1940).

/13/ Jean Paul Sartre, *Being and Nothingness*, trans. by Hazel Barnes (New York: Philosophical Library, 1956).

/14/ *Webster's Third New International Dictionary*, 3d ed., s.v. "anguish."

/15/ Martin Heidegger, *Sein und Zeit*, 6th ed. (Tübingen: Neomarius Verlag, 1947), p. 37. Note: the present author's own translation.

CHAPTER I

ANGST AND THE TRADITION

> The meaning of Being can never be contrasted with beings or Being as the abiding *ground* of beings; for "ground" becomes accessible only as meaning, and in itself is the abyss of meaninglessness.
>
> —Martin Heidegger, 1927

The Question at Hand

The purpose of this chapter is to sketch the general hermeneutical horizons within which Angst's role in Western thought may be interpreted. Such a task is, however, somewhat problematic, since in the literature few specific references to Angst are provided. An exception is J. Boehme's tangential references to Angst in his work of 1610, entitled *Aurora* or *Die Morgenröth im Aufgang.*/1/ Be that as it may, Angst does not emerge as a specific theme for philosophical investigation until well over two hundred years later in S. Kierkegaard's *Begrebnet Angest*/2/ of 1844, a work that can be traced indirectly back to Boehme's influence through the writings of Von Baader, Hamann, and Schelling.

This general lack of literature on Angst may be due in substantial part to a specific lack of interest in distinguishing Angst from fear. Another reason may be that the term Angst is neither a product of "rationalism" nor, for that matter, even Greek. Nonetheless, a review of the existing Angst literature shows that prior to Kierkegaard the distinction between fear and Angst had been blurred in the Western tradition. Therefore, the vague idea of Fear/Angst has been dismissed by philosophers as either anti-rational and therefore an anathema to philosophy; or some kind of passionate heresy that has no place in the pristine ivory towers of rationalistic thought.

What this all means is that underpinning the hermeneutic phenomenological conception of Angst is an elusive and somewhat risky business. Nonetheless, to accomplish our task we must not shrink from the tradition of Western philosophy; rather, we must confront it both cautiously and selectively—cautiously because no forced interpretation will disclose Angst in the tradition, and selectively because one cannot possibly consider the role of Angst in the entire history of Western thought in the scope of this essay. The problem, therefore, is appropriate as a suitable point of departure for this investigation. To this end hermeneutic phenomenology, the "methodology" of this essay, requires a return to the original concept of Angst itself as the means whereby a "first cut" or provisional understanding of Angst is made possible. Thus, the etymology of the term "Angst" may be helpful in fixing our departure point.

Etymology of Angst

According to the lexicon of "Indo-European Roots" in *The American Heritage Dictionary*, the word "Angst" comes from the Indo-European stem, *angh-* meaning roughly "painful, tight, or painfully constricted."/3/ Tracing Angst backwards from the present to Greek times, the following list reveals the conceptual integrity of *angh-* throughout the classical and Old High German linguistic traditions:

> Suffixed from *angh-os-ti* became in German Angst and in Old High German *Angest*.
> Suffixed from *angh-os-to* became in Latin *angustus*, meaning "narrow."
> The root *angh-* became in Latin *angere*, meaning "to strangle or draw tight."
> The root *angh-* became in Greek *anchine*, meaning "to squeeze."
> The root *angh-* became in Greek *anchone*, meaning "a strangling."/4/

This list is by no means exhaustive. The etymology of Angst likewise appears in the Greek concept of *Ananke*, "the force of binding necessity," an additional connection not appearing in the special lexicon cited above. Nevertheless the evidence is overwhelmingly in favor of the close knit relationship between Angst and *Ananke*, a relationship vividly revealed by the American depth-psychologist James Hillman,/5/ but one suggested earlier by Heinz Schreckenberg./6/ If these two thinkers are correct in their analyses, and there is no reason to suspect they are not, the etymological history of Angst goes far deeper than the Indo-European root *angh-*; indeed it delves deeply into prehistoric culture. On the basis of Hillman's insight, we may extend the root of Angst back into the ancient Near Eastern civilizations; beginning with contemporary Arabic *iznak* and ending at the founding concept of ancient Egyptian *hnk* which, like the Latin *angustus*, also means "narrow." (Please see the joint etymology of Angst and Ananke in Table 1 below.)

Hillman observes that Schreckenberg's evidence for the Near Eastern connection between Ananke and *hnk* ". . . extends far further than this digest and it tallies with the more usual etymologies of *Ananke*, relating it with the German *eng* (narrow), with *angina*, *angst* and anxiety, and *agchein* (Greek) to strangle, and with *agham* (Sanskrit, evil)."/7/

What is revealed about Angst in these etymological excursions into the murky and remote dawn of human history? First, early man experienced Angst as an intense physical sensation rather than a self-conscious feeling. This is not hard to understand when we see that no clear evidence exists to suggest "self-consciousness" in primitive man. Second, Angst was experienced by ancient man as a binding force; later as a metaphysical force which became for the Greeks Ananke, the goddess of necessity: she who binds the gods and men alike, and who with Chronos, the god of time, keeps the Kosmos together.

TABLE 1

JOINT ETYMOLOGY OF ANGST AND ANANKE

Language	*Word*	*Meaning*
Arabic	iznak	the cord that binds yoked oxen
Arabic	hannaka	necklace
Arabic	hanaqu	strangle
Chaldean	hanakin	fetters lain on the necks of prisoners
Hebrew	anak	chain formed necklace
Syriac	ḥnḳ	chain, suffocation
Akkadian	hanaqu	constrict, strangle, to wind tightly around the neck, as the band of a slave
Coptic	chalak	ring
Old Egyptian	enek	surround, embrace, strangle

Old Egyptian	hng	throat
Old Egyptian	ḥnḳ	narrow

SOURCE: Heinz Schreckenberg, *Ananke—Untersuchungen zur Geschichte des Worgebrauchs, Zetemata*, Heft 36 (Muenchen: Beck, 1964), pp. 169–74, as cited in James Hillman, "On the Necessity of Abnormal Psychology," *Eranos* 43 (Leiden: Brill, 1947), p. 97.

But all this is preliminary and serves only as a "first cut" on an explicit interpretation of Angst. This phenomenon is revealed throughout the Near Eastern, Greek, Roman, Old High German, and Modern German traditions, but it alters the face that it shows. From primitive terror of death by strangulation to the subtle and complex existential and phenomenological interpretations of Angst in post World-War II philosophy, Angst turns around a central theme: a pre-reflective apprehension of how human existence (Dasein) is restricted or thwarted by Being in general (*Sein*). To couch this in a theological context that dominates all but the very tip of the Western tradition of thought, Angst is, then, the pre-reflective apprehension of the abyss between God and man. In the following section, we shall ground this interpretation by examining Angst in mythopoeic thought.

Angst and Mythopoeic Thought

It can be argued, successfully one would think, that prehistoric man is not yet Dasein. Why is this so? For one thing, primitive man did not yet possess an explicit concept of 'World' as the transcendental totality of beings which, as a whole, forms a transcendental context of meaning. For another, primitive man related to his environment in a magical way rather than one governed by explicit onto-cosmological guidelines for understanding and interpreting existence in a World. What we will argue here is that the concept of World comes into being through the medium of "mythopoeic thought"—defined as that kind of thinking which speculates on the origin and the *telos* of the Cosmos, a thinking expressed in the poetic myths of creation. The meaning and significance of creation myth from a hermeneutic perspective is that creation myths represent a first effort to articulate (*Rede*) a new and fundamentally ontological understanding (*Verstehen*) of Being. This mythopoeic understanding is grounded, we believe, in an enormously powerful but pre-reflective apprehension

(*Grundbefindlichkeit*) that reveals to primitive man a primordial Chaos logically preceding Cosmos. As we shall see, this primordial Chaos is universally schematized in the West as the Abyss metaphor.

This pre-reflective apprehension of meaninglessness behind the lived-world is what we call "primordial Angst." It may be further described as an apprehension fascinating and daunting: fascinating because the enormous power and intensity of Angst and daunting because of the terror Angst engenders. Primordial Angst, then, is a pre-reflective, pre-rational apprehension of an absolute groundlessness of Being expressed metaphorically as an Abyss—a feeling that (1) there may be no meaning to existence in the lived world of experience; (2) there may be no transcendental unity to experience; and (3) there may be no transcendental meaning to the context abiding behind and beyond primitive man's lived experience. This apprehension, as we hope to show in the following pages, reveals many faces throughout the development of Western thought. It is the purpose of this chapter to discover them in a quasi-historical fashion, and to show how primordial Angst is at the center of each. Let us begin, with a look at our ground phenomenon, primordial Angst.

Primitive Man and His Environing World

The texts of the ancient Near East show beyond doubt that primitive man did not "think" or reason as does modern "scientific man." The evidence suggests, for example, that primitive man did not use as a conceptual tool anything like an independent principle of causality. Likewise, he was not capable of formally distinguishing between reality and appearance, between subject and object, or between internal and external use of time and space as we, modern and thoroughly scientific Dasein, understand these concepts./8/ Rather, primitive man appears to have enjoyed a special magical relationship with his environment, often substituting the conceptual for the perceptual, often seeing personality and will in what we call "inanimate objects." In fact, this appears to be the root of primitive animism: a direct intuition of an "I-Thou" relationship between ancient man and the "objects" of his experience, such as stones, trees, animals, and the wind. For primitive man each of these, along with the potent natural forces of the Earth, was imbued with a unique personality and will of its own. No one has described this special relationship better than Henri Frankfort, who observes:

> The world appears to primitive man neither inanimate nor empty but redundant with life; and life has individuality, in man and beast and plant, and in every phenomenon which confronts man—the thunderclap, the sudden shower, the eerie and unknown clearing in the wood, the stone which suddenly hurts him when he stumbles while on a hunting trip. Any phenomenon may at any time face him, not as "It" but as "Thou." In this confrontation, "Thou" reveals its individuality, its qualities, its will. "Thou" is not contemplated with intellectual detachment, it is expressed as life confronting life, involving every faculty of man in a reciprocal relationship. Thoughts, no less than acts and feelings, are subordinated to this experience./9/

When the special relationship between man and his environment is disrupted or uprooted by the operations of inexplicable "natural" forces which defy understanding and interpretation, primordial Angst is first revealed. Primitive man, who in abject terror quakes in the face of a thunder and lightning storm that sets the trees ablaze, who watches in anger and frustration as the flood carries off his mate and offspring, who gazes with horror at the leprosy that emasculates his own flesh, and who can do absolutely nothing about these experiences; this man begins to know Angst in its most terrible and dark face. For it is not so much these powerful forces that give rise to his terror and awe; rather these grow from a primitive urging to question *why are these things happening*? and by far more important, *why are they happening to us*? Primordial Angst, then, is for such a man a pre-reflective apprehension that something is wrong, or out of place in the way that "things should work" in the magical relationships which form an "I-Thou" animistic meaning-context whereby primitive man interprets his lifeworld. It is in the awe-filled mystery of "Why?" that first reveals primordial Angst, because to that question there is never a satisfactory answer. Thus, in primordial Angst the first attempts at an answer are made possible by an overwhelming demand for justification and meaning to the chaotic experience of being human.

As we argued above, the answers to such questions take the form of *mythopoeic thought*, the means whereby primitive man transcends the naivete of animism and emerges into a Cosmos.

Cosmos may be seen as a new conceptual order wherein man becomes Dasein—a being who is set over against Being as the transcendental totality of all beings dwelling in a contextual and referential World. The significance and meaning of World as Cosmos is revealed to the new Dasein in an elaborate but sometimes obscure system of mutual interconnections. If this is so we must now how mythopoeic thought was carried about in those different civilizations which have profoundly influenced our Western traditions. To do this we need to look at their respective creation myths.

The Creation Myths of the Ancient Near East

In Mesopotamia at the dawn of history (ca. 5000 to 3500 B.C.E.), an early creation myth typical of mythopoeic thought states: in the beginning there was only Nammu, the great goddess-mother and the oceanic abyss. Nammu was conceived as a vast womblike Nothingness from which all beings emerged./10/ In a much later Babylonian version of the creation (ca. second millennium B.C.E.) entitled the "Enuma Elish," the abysmal mother is reinterpreted as vengeful Tia'mat who must be slain by Marduk, god of Babylon, so that her corpse can be split into the dome of sky above and the earth below, to form a Cosmos./11/ It is not incidental that in this second creation myth, Marduk fashions men to serve the gods as though by afterthought,/12/ for in so doing, he forever places Babylonian Dasein in the strangling yoke of slavery to the gods.

In Egypt an entirely different set of circumstances obtained. In the beginning, according to myth, there was only Nun, the father of all Being, conceived as a watery abyss./13/ From Nun arose Atum, the primeval hillock identified by the ancient Egyptians as their creator god. Atum, in turn, created the Ennead of the Egyptian pantheon: Shu-Tefnut, Geb-Nut, Osiris-Isis and Seth-Nephthys./14/ In the Egyptian account, there is also identified the primeval Ogdoad, or "Eight Weird Creatures" that make up the watery Abyss. The Ogdoad was, as we shall see, to impact profoundly the later biblical account of creation in Genesis. The Ogdoad consisted of: Nun, the abysmal waters and his consort, Naunet; Kuk, the darkness and his consort Kauket; Huh, the boundless stretches of the formless and his consort, Hauket;

and finally Amun, the hidden intangibility of the abyss and his consort, Amunet./15/

In the Hebrew creation myth at least three of these Egyptian Ogdoad reappear as the primordial condition present before the creation of the universe by Yahweh. "In the beginning of creation," the Book of Genesis begins, "when God made heaven and earth, the earth was without form and void, with darkness over the face of the abyss. . . ."/16/ The King James version translates the Hebrew phrase *hosek al-penai tehom* as "darkness was upon the face of the deep." But as Gerhard Von Rad has suggested, the word for the deep, "tehom" in Hebrew, is unquestionably related to Tia'mat,/17/ the Babylonian abyss mentioned above. Likewise the darkness referred in the Hebrew version, "*hosek al-penai tehom,*" *hosek* is a clear echo of Kuk, the Egyptian primordial darkness.

In the Greek tradition, Homer was acutely aware of the abysmal waters of Okeanos, "from whom the gods are sprung."/18/ But in Hesiod's *Theogony*, the abyss metaphor is replaced by airy chaos, conceived by Hesiod as a yawning gap ". . . that is filled with heat when Zeus makes his thunderbolts."/19/ Along with Chaos, Earth and Eros form the primordial trinity that was present before creation. Thus, Chaos in very ancient Greek mythology is neither a watery abyss nor a formless disorder, but rather the yawning gap between the dome of heaven and earth.

A second Greek account of creation is given in Plato's *Timaeus*, but we shall reserve our discussion of Plato's creation myth for a later section of this chapter. Suffice it to say for the present that for Plato, Chaos was indeed envisioned as a primal and disorderly movement, but not specifically as an abyss./20/

Early Dasein's World

What is revealed in these respective creation myths? Clearly they reveal a newly emerging Dasein which finds itself standing within a Cosmos-Ontos that is self-created. Specifically, as we saw above, in Mesopotamia and Babylonia, Dasein became secure through mythopoeic thought by becoming a slave to the whims of the gods, thus finding a place in the Cosmos. The Cosmos was, in other words, envisioned as a limitless political state upon which Mesopotamian Dasein could depend. Dasein's primary virtue, absolute obedience to the Gods, was well defined and well understood. What was the reward for such obedience? It was the protection of the gods and their rewards and favors in the form of health, wealth, long life, honor, and perhaps more important, many sons./21/

Egyptian Dasein appeared secure in the youthful vigor of Egypt's old Kingdom. The life giving waters of the Nile delta teeming with lush greenery amidst the baked and arid desert instilled a sense of confidence in life for the early Egyptians. The Nile's life-affirming waters demonstrated to Egyptian Dasein that behind the scenes Ma'at was functioning as the guarantor of Egypt's abundance. Ma'at, schematized as an Egyptian goddess, was the principle of justice or truth,/22/ a principle, moreover, applying equally to gods and goddesses as well as humans through the medium of the Pharoah. Belief in Ma'at's power extended throughout the roughly forty centuries of Egypt's glory. Hence, the role of Dasein in the Egyptian cosmos was to conduct its life in accordance with the principle of Ma'at, for without such a principle, the abyss of Nun would surely return. "Ma'at," says Von Rad, "guarantees the continuance of world, of both the cosmic world and the social world of men. Gods and men live by it."/23/ Dasein's place in the Egyptian Cosmos was not only to conduct its life in accordance with Ma'at, but to ". . . translate it into reality and hand it on."/24/ Ma'at therefore is Egyptian Dasein's solution to the daunting apprehension of primordial Angst. By capturing Being in the web of Ma'at, Egyptian Dasein was redeemed from the terrors and awe of the primal forces and ruthless "barbarian" people surrounding Egypt. Belief in this primary universal force served to cover over the more primitive primordial Angst engendered by the apprehension of Nun's dark abyss.

In the ancient Hebrew tradition, the power and personality of Yahweh spelled out clearly the proper role of Hebrew Dasein in a monotheistc conception of Being. Yahweh's original covenant with Israel defined for the first time a specific *ontic role* of Dasein: the Ten Commandments. Dasein's ontological role, however, was founded in the notion of *fearful reverence*. Thus, a direct personal commitment of faith in God or a reverential "fear of the Lord" provided the ontological ground for Hebrew Dasein's conception of Being. If Hebrew Dasein's primordial Angst was lessened by the Yahweh-Israel covenant, it was clearly never eradicated. Indeed, Angst in the face of the Abyss is most visible in the wisdom and prophetic literature of the Old Testament, surfacing in the synoptic gospels of the New Testament.

As for the Greeks, we must remember that the so-called "pre-Socratic" thinkers, Thales, Anaxamander, Anaximenes et al., were utterly unique in the tradition of Greek thought even during their own lifetimes. The pre-Socratics' attempts to account for change without recourse to a deity is precisely what makes their contribution to Western thought so valuable. This Ionian proto-scientific account of change coupled with a free spirit of inquiry distinguishes these early lovers of wisdom from their contemporaries, and specifically from the earlier cult religion centering around the Bull-god, Dionysus.

Nonetheless, the Dionysian cults were a more primordial reflection of ontological Angst. Dionysian cultism was brought to Boetia by Thracian tribes migrating from their homeland. The cult of Dionysus spread rapidly throughout Attica, to the Aegean Islands, and finally to Peloponnesus itself. The essence of their rituals centered around the unique concept of "orgy," a term originally meaning only "act of ritual," and a word specifically used in connection with the Eleusian mysteries and the Dionysian omophagia./25/ The descriptions of such orgies can hardly be interpreted from a philosophical perspective as simple degeneracy or insanity. Rather, since survival depended upon nature's moods, Greek Dasein celebrated the birth of spring and mourned the death of fall. These rituals were, in short, the *praxis* of mythopoeic thought, whereby primordial Angst was assuaged through questing for union with the divine.

The excesses of these cults were soon tempered by the Homeric Greeks; that is, the cannibalism and sexual excesses of the Dionysian cults were checked and rechanneled into more ritualistic, cerebral practices which led to Western mysticism. The Dionysian cults became merged with the Eleusian agricultural rites. Ultimately, by the seventh century B.C.E., they blended with Orphism to form a unique religion based on a combination of three notions: original sin, divinity of the soul, and transmigration.

The frenzy of cultism, with its quest for union with the divine, reveals the pre-reflective apprehension of the abyss. It was only later that Angst became devitalized in Orphism, but originally it was the potent foundation upon which Greek mystery cults stood. Primordial Angst again was the condition necessary for the frenzied orgies, ritual sacrifices, drunkenness, sexual abandonment, and bestial dances of the Dionysians. In such practices, Dasein was freed from ontic concerns to merge with Being or nature's essence in the form of Dionysus the bull-god. Hence, these were colorful and primitive quests for union with the divine, an ecstatic union which established a firm foundation for the mystical tradition of the West, a tradition that became the means for disclosing Angst's meaning from the time of Augustus to Jacob Boehme.

Our point in this discussion has not been to minimize the role of the so-called pre-Socratic philosophers in the Greek tradition. Rather we have tried to point out that contemporary philosophical thinkers often tend to obscure the importance of the mystery cults of ancient Greece. Clearly even the proto-scientific gropings of Thales can be interpreted as being imbued with primordial Angst. After all, the three doctrines of Thales, namely: (1) water is the nature of all things; (2) all things are possessed with a soul; and (3) all things are filled with *daemons* or *gods*,/26/ imply the presence of mythopoeic thought par excellence, namely, *the gods cause motion*. Surely this is a notion firmly grounded in the primordial Angst tradition. Why? Because such Angst is a primordial predisposition of other Ionians as well. Behind their attempts to explain the nature and operations of the cosmos, there lurks the realization that to explain change is in some sense to rob it of its power over Dasein. So, when Aristotle ignores the second and third propositions of Thales, and focuses only on Hylozoism, he covers over the glimmer of primordial Angst revealed in Thales' thought; namely the early doctrine of the *soul and of demons to explain motion*./27/ Something, after all, must explain change.

Thus, for early Greek Dasein, the cult practices captured Being, allowing Dasein, if only for a flickering moment, to participate in Being, schematized as the god, Dionysus. This meant that Dasein's place was one of worship, sacrifice, and orgiastic ritualistic participation to quell the primordial Angst that loomed close to the surface of existence in such a world-view.

Angst and the Near Eastern Wisdom Literature

The cosmologies we have discussed in the previous section, those of the Mesopotamian, Egyptian, Hebrew, and Greek civilizations, are each self-contained regarding their respective ontological assumptions concerning the Cosmos. But in the Near Eastern texts as a whole there is a jointly shared theme disclosing another face of primordial Angst. Our task in this section is to discover this new face which we shall designate as "onto-theological Angst," to distinguish it descriptively from "primordial Angst."

Onto-theological Angst can occur only when the *Ontos-cosmos*, or the transcendental totality of all beings, is interpreted as a Supreme Being. This Ontos-cosmos then becomes schematized as a pantheon of deities or a monotheistic God. Onto-theological Angst, therefore, refers to a *pre-reflective apprehension* of a disrelationship between Being and beings, interpreted as a breakdown in relations between "Deity" and the "creatures of god." Apprehension arises as a source of objective uncertainty regarding the relationship between man and God, since prior to the apprehension the positive relationship was certain and secure. The pre-conditioned positive relationship appears to have been in priestly dogma that grew out of the creation myths which gave meaning to the life-world of Dasein. When the apprehension surfaces, Dasein is fundamentally uprooted from the gods or God as the transcendental ground or source of the lived world. Dasein is left, therefore, in the face of an abyss of meaninglessness.

To ground our discussion of onto-theological Angst in the shared wisdom literature of the near East we will discuss Angst from the viewpoint of a universal theme shared by the Mesopotamian, Egyptian, and Hebrew traditions: namely, the theme of innocent suffering, sometimes called "the Joban theme."

The earliest version of the Joban theme is found in an ancient Sumerian text written ca. 6000–5500 B.C.E.. W. G. Lambert, translator of the "Mina-Arni" texts, shows that the problem of innocent suffering may begin with Mina-Arni stating: "I have been treated as one who has committed a sin against his God."/28/ The theme is carried further in S. N. Kramer's translation of "Man and His God,"/29/ wherein a formerly rich, wise, and upright man is now tormented with sickness and anguish for what appears to have no justification. But adhering to the Mesopotamian belief that no man is free of sin, the protagonist accepts his fate, prays for deliverance, and ultimately receives deliverance, again with no justification or reason provided by Deity. In yet a later version, entitled, "I Will Praise the Lord of Wisdom," also known as "The Babylonian Job,"/30/ a morally upright man is abandoned by his friends and his god. In the depths of onto-theological Angst, he cries out:

> I gave my attention to supplication and prayer; sacrifice was my rule; the day for reverencing the god was a joy to my heart! . . . I wish I knew that these things were pleasing to one's god! . . . Who knows the will of the god's in heaven?/31/

A final example of the Mesopotamian versions of the Joban theme is entitled, "The Babylonian Theodicy."/32/ In this work, closest in style and format to the Book of Job, the protagonist is once again an innocent sufferer comforted by a friend. Bemoaning his unjust suffering, the protagonist pleads with onto-theological Angst, "Can a life of bliss be assured? I wish I knew."/33/ His tradition-bound comrade tells him to have patience and hold firmly to his faith; for, "the way of the gods is remote, . . . knowledge of it difficult."/34/

In a third-millenium B.C.E. Egyptian text entitled, "A Dispute over Suicide,"/35/ a man debates with his soul as to the expediency of suicide when life becomes impossible: again onto-theological Angst is revealed as the protagonist laments:

> To whom can I speak today?
> I am laden with wretchedness
> For lack of an intimate (friend).
> To whom can I speak today?
> The sun which treads the earth,
> It has no end./36/

Finally, there is the Old Testament Book of Job itself. The story of Job's suffering; of the attempts of his friends, Eliphaz, Bilbad, Zophar, and Elihu to advise and console him; of Job's demand for Elohim to justify himself to Job; of God's awesome questioning that ultimately convicts Job; and of Job's return to the grace of God: these themes are well known and require no elaboration here. What *is* important to discuss here is the place of onto-theological Angst in the Book of Job. From our point of view such Angst is disclosed as the primary pre-disposition of the entire poem. How is this so?

Reduced to its essence, the Joban theme can be posed in the form of questions from the righteous sufferer to his Deity which might proceed along the following lines: Why must I suffer when I have lived in accordance with your divine commandments? Why must I suffer when I am therefore guilty of no sin against you? Why must I suffer when I have trusted in you completely, held you in reverence, and maintained my faith in you as the Creator and sustainer of my being and of Being itself? Surely, these questions are no mere request for information. They go far deeper than that into the grounds of a creature creator relationship—a relationship originally established in the mythopoeic thought which gave rise to creation myths in the first place. At root, then, these questions are more of a *demand* by the creature that the creator *justify*—not simply explain—why the righteous suffer, to demand a divine assurance that the intentions of God are, at least in the case of Israel, consistent with the Old Covenant. Archetypal Job, then, is Dasein thrown into a world of un-interpretable suffering. Further, this Dasein, fallen from the Grace of Being, must endure the pregnant silence of God's voice, hoping for deliverance from a terrible fate, one literally worse than death. But what is this terrible fate? Surely it is not just the anguish of physical suffering. Indeed, Job has shown that such anguish can be endured. Rather it is onto-theological Angst, the Angst in the still voice of Elohim's silence, from which Job seeks deliverance.

And what does God's silence reveal? It reveals a heretofore unsuspected abyss existing between creator and creature, between archetypal Job and his god or gods. But at root what Job experiences in God's silence is that it is *impossible* for Dasein to understand fully and interpret God as the transcendental totality of Being—a realization that jars Dasein and uproots it from its smug grasp of God *qua* Being as such. In Job's case, as a direct result of his onto-theological Angst, he demands a face-to-face confrontation with Elohim.

> Terror upon terror overwhelms me
> it sweeps away my resolution like the wind
> and my hope of victory vanishes like a cloud. . . .
> I call for help, but thou does not answer;
> I stand up to plead, but thou sittest aloof
> thou has turned cruelly against me
> and with thy strong hand pursuest me in hatred./37/

To be sure, these are strong words; words possible, we suggest, only because of the abysmal depth of Job's onto-theological Angst. Such Angst is clearly no mere ontic feeling of "anxiety," "dread," or "anguish." This Angst is a basic pre-disposition toward the ground of Job's being, or perhaps more to the point, toward the absence (silence) of the ground of Job's being. Thus, Job's onto-theological Angst is an immediate but objectively uncertain recognition of an infinite abyss that separates Dasein (Job) from *Sein* (Elohim). This recognition gives the Joban theme, the theme of innocent suffering, its profoundly touching point, power, and universality.

This theme, and the onto-theological Angst associated with it, are carried over into Deutero-Isaiah where innocent suffering emerges again as the theme of the Suffering Servant. Later the same theme reappears in the synoptic gospels of the New Testament as the theme of the Suffering Savior. Hence, the universal suffering and the Angst revealed in it provides a major theological and ontological connection point between the Old and New Testaments.

Angst and Ananke—The Greek Contribution

To begin the discussion of Angst and Greek thought it is necessary to backtrack as promised to Plato's account of the creation of the *Timaeus*; for in this work is disclosed another facet of the Angst phenomenon which we shall call "strictural Angst," to distinguish it descriptively from the "primordial Angst" and the "onto-theological Angst" of the previous sections. Strictural Angst, as a face of primordial Angst, is revealed as a pre-reflective apprehension of the ontological restrictions placed on Dasein's will, a notion schematized by the Greeks as the goddess Ananke.

Turning, then, to the creation myth in the *Timaeus*, Plato argues that the ordering of the Cosmos from primordial Chaos was efficiently caused by the Demiurge. This could not have occurred, however, without the aid of two independent first principles who in their personified forms were independent from the Demiurge. Plato says:

> Now the foregoing discourse . . . has set forth the works wrought by the craftsmanship of Reason [*Nous*]; but we must now set beside them the things that come about by necessity [*Ananke*]. For the generation of this universe was a mixed result of the combination of Necessity *by persuading her* to guide the greatest part of the things that become towards what is best; in that way and on that principle this universe was fashioned in the beginning by the victory of *reasonable persuasion over Necessity*./38/ (Italics added.)

As we argued at the outset of this section, strong evidence

has been offered for an etymological connection between Angst and Ananke, the latter interpreted by the Greeks as "necessity." In the *Timaeus* passage cited above, Ananke appears as the *errant* cause of creation. Ananke, then, is a principle the Greeks associated with rambling, digressing, irrational, irresponsible, deviating, irregular, and random./39/ For Plato, Ananke is a true *arche*, that is, a "first principle" not derived or derivable from anything else. Its role as errant cause in the creation myth is, we propose, a mythopoeic schematization of strictural Angst, the Angst experienced in the pre-reflective apprehension of limits to reason, and more specifically, to freedom of the will. This view finds confirmation in Plato's depiction of Ananke given in the *Cratylus*. There Plato observes that the idea of Ananke ". . . is taken from walking through a ravine which is impassable, and rugged, and overgrown, and impedes motion—and thus is the derivation of the word necessity."/40/

The Character of Ananke in Greek Thought

But the conception of Ananke as binding necessity, of "no way out" of a given situation, predates Plato's writings and goes back to Orphic and Pythagorean mythology where Ananke was mated with the great serpent Chronos. Thus, Ananke (necessity), and Chronos (time), form a binding ring encircling the entire Cosmos, limiting the possibilities of our intellect, will, and desires. One cannot exist without the other. The results of this union for Dasein are nowhere better described than by Hillman.

> Time and Necessity set limit to all the possibilities of our outward extension, of our worldly reaches. Together they form a syzygy, an archetypal pair, inherently related, so that where one is the other is too. When we are under the compulsion of necessity, we experience it in terms of time, e.g., the chronic complaints, the repeated return of the same enclosing and fettering complexes, the anxiety occasioned by the shortness of our days, our daily duties, our "deadlines."/41/

This insightful observation discloses the role of anxiety in Greek thought, but it only hints at a much deeper role of primordial Angst as anxiety's ontological pre-condition. Strictural Angst appears here as the primordial pre-disposition (*Grundbefindlichkeit*) of Greek Dasein. Only after Angst's first primordial appearance can strictural Angst be interpreted and understood through mythopoeic schematization and personification as Ananke. Ananke is mythologically interpreted originally as a goddess but later became the philosophical idea of necessity in Aristotle's thought. As an idea, Angst/Ananke is interpreted as an errant cause that Dasein recognizes in the irrational, the irresponsible, or the indirect, and which possesses the attributes of rambling, digressing, etc., discussed above. In the soul (*psyche*) Ananke becomes the producer of irrationality and frustration. In the broader context of the Cosmos, Ananke becomes the dialectical opposite of *Nous*, of which the Cosmos is then an admixture. The absolute and literal centrality of Ananke is revealed in R. B. Owens' "The Knees of the Gods,"/42/ where the Goddess Ananke is interpreted as the spindle of the universe. She governs the movement of the soul and its freedom as well as the motion of the stars. Ananke, then, as the personification of Angst, is the "still point of the turning universe" for the Greeks, an irrational Goddess who must be persuaded by *Nous* to "guide the greatest part of things that become toward what is best."

The philosophical idea of necessity was stripped of Ananke's irrational power by the time of Aristotle's *Metaphysics*. For Aristotle Ananke is an internal principle identified as a function of the true nature of things, defined as "that which impedes and tends to hinder, contrary to purpose."/43/ In Hillman's view, this bland, mechanical, and leveled interpretation of Ananke determined our thinking about Ananke ever since./44/ Aristotle's Ananke interpretation is a perfect example of how the power of Angst has been systematically covered over by the Western metaphysical tradition—an indictment frequently made by Heidegger, but not, so far as we know, in this instance.

Ananke and the Tragedians

This leveled view of Ananke was certainly not appropriated by the Greek tragedians, the contemporaries and in many ways the competitors of the Greek philosophers. For the tragedians, Ananke retains all the power of strictural Angst. Consider, for example, the plaintive cry of Aeschylus' bound Prometheus:

> Oh Woe is me!
> I groan for the present sorrow
> I groan for the sorrow to come, I groan

> questioning where there shall come a time
> when He shall ordain a limit to my sufferings.
> What am I saying? I have known all before,
> all that shall be, and clearly known; to me,
> nothing that hurts shall come with a new face.
> So much I bear, as lightly as I can,
> Destiny that fate has given me;
> for I know well against Necessity [Ananke],
> against its strength, no one can fight and win./45/

Like Job, Prometheus does not suffer simply from physical anguish, rather he suffers from the Angst of his existential situation. But unlike Job, Prometheus fully understands that situation as the restriction of his freedom, and his will.

Euripides, in his *Alcestis*, defines the aloofness of the goddess Ananke: "She alone is goddess without altar or image to pray before. She needs no sacrifice."/46/ Ananke as a power is not one to which man can appeal. As a goddess she binds the freedom of both gods and men; and from her web, no escape is ever possible. Thus, the metaphors attached to the conception of Ananke's goddess are binding, encircling, cords, nooses, collars, knots, spindles, wreathes, harnesses, and yokes: all metaphors for her restrictive and constrictive aspects in a strictural context, the birth of which is the primordial pre-disposition of strictural Angst.

The Angst of Gnosis

In this section we hope to describe another facet of Angst which for the sake of clarity we shall call "pneumatical Angst" to distinguish it descriptively from the three facets discussed previously. It would be helpful to repeat here, however, that these facets are not mutually exclusive. Rather, they are but differing ways of interpreting the ground phenomenon of primordial Angst as the pre-reflective apprehension of the essential incompleteness, meaninglessness, and ambiguity of the life-world, revealed first in the terror and awe experienced by primitive man in the face of the abyss. These differing facets are, therefore, hermeneutical interpretations of primordial Angst, interpretations possible only within the context of a Cosmos as Ontos.

General Features of Gnosticism

Very generally speaking, the world of Hellenistic Dasein was permeated with uprootedness, change, value confusion, and a general ontic anxiety in the face of bustling confusion, especially at the early stages of the eclectic world view which resulted from Alexander's conquests. Likewise, of course, this was an exciting, robust period of vast spiritual change. But beneath the staccato lifestyle of cosmopolitanism, the dark current of ontological Angst moved forward as the primordial pre-disposition of Hellenistic Dasein. Dasein's horizons had considerably expanded by this time, thanks to a blending of the Near Eastern cultures on one hand and the Greco-Roman world-view on the other. More specifically, the characteristics marking the Hellenistic world-view wherein pneumatical Angst is revealed are: (1) the growth of Hellenized Judaism, with Philo of Alexandria as its chief spokesman; (2) the development and dissemination of Babylonian astrology, magic, and fatalism; (3) the flourishing of the Eastern mystery cults, such as Zoroastrianism, and the evolution of such cults into spiritual mystery religions; (4) the rise of Christianity to prominence and influence; (5) the general efflorescence of Gnosticism from Coptic and Hebrew sources; and (6) the upsurge of transcendental philosophical movements of late antiquity, beginning with Neo-pythagoreanism and culminating in Neo-platonism./47/

Against this background, Dasein was ontically a *cosmopolite*, that is, "a good citizen of the cosmos." Such citizenship was seen as a moral end for Dasein: Dasein was now in possession of the universal *logos* as a citizen of the Cosmos. This gave rise to an utterly new phase of being-in-the world, the concept of being-a-private-person or a "self"; an idea that was not possible in the tribalistic mentality of the Near East nor, for that matter, in the city-state mentality of Hellenic Greece. Dasein was no longer "of the polis" or "of the tribe," but rather "of the Cosmos" itself. As Dasein, man was freed to pursue his own ends; a big step indeed, as it makes possible for the first time the interpretation of a free-being whose allegiances are to himself first and to his community second. But as one can well imagine, such freedom fundamentally uprooted Dasein, alienating Hellenistic Dasein from the cultural traditions as well as the theological presuppositions which had previously provided a pre-reflective meaning-context for interpreting the flux of experience. Hellenistic Dasein was therefore forced to plow new ground for interpreting existence; for the new isolated world of uprooted traditions now seemed cold, hostile, and markedly alien.

The primordial pre-disposition accompanying this understanding of existence was, of course, Angst: a basic Angst that shows through the writing of Gnosticism in general. So, to understand the Angst of Hellenistic Dasein we must first understand something of the religio-philosophical movement known by the collective and generic name, "Gnosticism."

The English word "gnostic" comes from the Greek word *gnosis* meaning "revealed knowledge." Thus, gnosis is not to be confused with *episteme* or "discursive knowledge." Rather, gnosis has a religious, spiritual quality which aligns it more with faith than reason, the heart more than the mind. Specifically, gnosis means a revealed knowledge of a transmundane God. Nonetheless, gnosis was considered a practical form of knowledge through which man's essence, his spirit or *pneuma*, could attain salvation and return to the Godhead beyond what the gnostics regarded as the evil Cosmos. For gnostic thought, this world is not the creation of the ultimate, transmundane God. Rather, it was created by a malevolent Demiurge, sometimes associated with Yahweh, who was ignorant of the transmundane Godhead above his limited being. For the gnostics the Demiurge's malevolent ignorance of the true God accounts for evil and suffering in this world. From this perspective Dasein may be seen as a being thrown violently into an alien creation; that is, Dasein experiences the Cosmos as wholly alien to its pneumatic essence, but does not understand why this is so. Hell is in this world, not in the next, and from it there is no exit.

Throughout Gnosticism emerges a single motif concerning the plight of man: his alienness from this world and the Cosmos of its evil creator. Such alienness is interpreted by Hans Jonas/48/ and Susan Taubes/49/ as a form of metaphysical homesickness, a longing for a return to a home beyond this world where gnostic man truly belongs. Such a haunting longing is experienced as an ontic "uncanniness," a "not-being-at-home here," an ontic feeling made possible by pneumatical Angst. Nonetheless gnostic Dasein considered alienation to be a mark of spiritual excellence; for he who experiences it has had the call and has thus taken the first step on the way back to the Godhead as the very ground of his Being. In the literature of Mandaean Gnosticism, the pre-disposition of alienness is a root metaphor, one defining a basic characteristic of gnostic thought./50/

The Role of Pneumatical Angst in Gnosticism

The primordial pre-disposition of Angst, therefore, is the pre-interpretive ground of the gnostic world view. Gnostic Dasein experiences pneumatical Angst as the pre-reflective apprehension of Dasein's "being thrown" into an evil, alien world. Dasein thus experiences Angst as a dark and foreboding terror of the world itself, a place where estrangement and isolation is the permanent and "natural state" of Dasein's existence. As with Job's onto-theological Angst and Prometheus' strictural Angst, this is no mere ontic anxiety, dread, or anguish, although these ontic notions are clearly present. Rather, pneumatical Angst is a pre-reflective apprehension of the helplessness and homelessness of Dasein, of the isolation and deprivation of this world, of the bondage of man in this world, and the vacuity of human existence seen as a thrown-projection into a meaningless world. For gnostic Dasein no release was possible save through salvation from this evil existence. The Naassene Hymn observes of the gnostic spirit:

> Therefore, clothed in a watery form, she grieves, toy
> and slave of death.
> Sometimes, invested with royalty, she sees light.
> Sometimes, fallen into evil, she weeps.
> Sometimes she weeps and sometimes she rejoices;
> Sometimes she finally finds no exit, because her
> wandering ways have led to a labyrinth of evils./51/

Gnostic Dasein is a being-in-an-alien-world. Thus it is not surprising that Dasein interpreted this world as an evil place wherein Dasein is not at home. This interpretation is articulated in gnostic texts as alienness, homelessness, anxiety, and thrownness.

We could argue that pneumatical Angst is related therefore to primordial Angst inasmuch as the abyss, a metaphor frequently used in gnostic thought, is gnostic Dasein's daunting apprehension of the evil Cosmos. Consider the following evidence from the Nag Hammadi Library:

> The beings of his likeness, however, were exceedingly afraid. . . . Therefore, they fell down to the pit of ignorance which is called "the Outer Darkness" and "Chaos" and "Hades" and "the Abyss." He set up what was beneath the order of the beings of thought as it had become stronger than they. They were worthy of ruling over the unspeakable darkness, since it is theirs and is the lot which was assigned to them./52/

From this perspective, the evil world of gnostic man is interpreted as the abyss itself, an interpretation consistent with other features of the gnostic belief system. Thus, pneumatical Angst shows itself in its fascination form as the spiritual longing for salvation from an evil world of darkness and ignorance. In this vision of Being, the high God as the ontological ground of gnostic Dasein slips further away, beyond the very Cosmos itself to "another world," another realm of Being that is the true place where Dasein is finally at home. The ontological ground of the ontic or lived-world is Hell itself, interpreted as the abyss, a place into which gnostic man is thrown, is utterly alone, and is radically alienated from his true being.

Here is not the place to draw out the Heideggerian implications of this view, a point both Jonas/53/ and Taubes/54/ take seriously. It is only necessary to remark that the force of Gnosticism in general and its concept of the evilness of creation in particular have been transmitted by Augustine and the Apostolic Fathers as prime examples of heresy/55/ in early Christian times. Yet, Augustine was a Manichean Gnostic for nine years before becoming a Christian. His mysticism is therefore not due simply to his encounter with Neoplatonic thought, as we shall see in the next section. Rather Augustine's thought is likewise related to the deep mystical strain in Gnosticism; for gnostic thought contains the seeds for much of the Christian mystical tradition in the West. Notions such as the divine spark of man's spirit, the Godhead above God, and the yearning for union with the Godhead are the very essence of Eckhart's thought. These notions also appear in the Friends of God movement whose mystical work, the *Theologica Germanica*, was to profoundly influence Luther and the Protestant reformation around him. There is, then, an unbroken line from the pneumatical Angst of Gnosticism to the mystical or what we shall call "passional Angst" of the third century and beyond. To show this, we turn now to the place of Angst in the mystical tradition of the West.

Angst and the Mystical Tradition

Since the discovery of Angst in the mystical tradition is the purpose of this section, we shall designate the mystical facet of primordial Angst as "passional Angst to distinguish it descriptively from the other faces of the ground phenomenon of primordial Angst discussed above.

Passional Angst may be described as a pre-reflective apprehension of the uselessness of human ontic self-will in passionately seeking to attain union with ultimate Being. In other words, passional Angst is the apprehension of a profound barrier between the soul and the Godhead, a barrier resulting from our soul's willfulness, or passionate craving for union with the Godhead.

Since Kierkegaard's analysis of Angst is partially grounded in Augustine's reflections on ontological anxiety, we must begin with Augustine and passional Angst.

St. Augustine

The *Confessions* of St. Augustine are the clearest sourcebook of his personal mysticism in general as well as his passional Angst in particular. Regarding Augustine's general mysticism, Book VIII of the *Confession* is a *locus classicus.* While the tale of Augustine's conversion is familiar to many, it may be worth repeating here in his own words, for Augustine clearly shows passional Angst to be the condition immediately precipitating his conversion:

> But when a deep consideration had from the secret bottom of my soul drawn together and heaped up all my misery in the sight of my heart; there arose a mighty storm, bringing a mighty shower of tears. . . . I cast myself down, I know not how, under a fig tree, giving full vent to my tears; and the floods of mine eyes gushed out, an *acceptable sacrifice to Thee.* And, not indeed in these words, yet to this purpose, spake I much unto Thee: *And Thou, O Lord, how long? how long, Lord wilt Thou be angry, for ever? Remember not our former iniquities,* for I felt that I was held by them. I sent up these sorrowful words; How long? how long? "tomorrow, and tomorrow?" Why not now? Why not is there this hour an end to my uncleanness?/56/

The story continues with Augustine hearing the voice of a child from a neighbor's house saying, "Take up and read; take up and read." Augustine, believing this to be a command from God, opened the New Testament at random to Romans 13:13–14.

> I seized, opened and in silence read that section, on which my eyes fell: *Not in rioting and drunkeness, not*

> *in chambering and wantonness, not in strife and envying: but put ye on the Lord Jesus* Christ, and make not provision for the flesh, in concupiscience. No further would I read; nor needed I; for instantly at the end of this sentence, by a light as it were of serenity infused into my heart, all the darkness of doubt vanished away./57/

The passional Angst revealed behind this confession is striking. Augustine interpreted his present existence as a state of willful sin. He desired, indeed, passionately craved, salvation from sin, and experienced the passional Angst that precedes and paves the way for union with the One. Augustine understood his relationship to God only because of the Angst he felt in the face of a self-willed life of debauchery. His decision to pray in his unique manner, "how long wilt Thou be angry," reveals passional Angst concerning the disrelationship between God and himself.

Augustine's mature understanding of passional Angst, however, is shown in an important distinction he draws between "chaste" and "servile" fear, one articulated in his *Enarrationes in Psalmos.*/58/ We should not expect Augustine's distinction to be between "chaste Angst" on the one hand and "servile fear" on the other because, as we indicated in the opening pages of this chapter, the formal distinction between Angst and fear begins with Kierkegaard.

Augustine's description of servile fear is comparable to what we described in our Introduction as ontic "dread." Servile fear is ontic rather than ontological. It is a dread of circumstances that can cause alarm, worry, and disquiet. For example, servile fear may be experienced in a long-term illness, or in the threat of having to go to prison, or in losing one's reputation. Augustine likens "servile fear" to the feelings of an adulterous wife who has an anxiety that her husband might discover her secret affair, or worse, that she might be caught by her husband in the arms of her lover.

"Chaste fear," on the other hand, is ontological. Properly speaking it is passional Angst—the primordial pre-disposition which transcends the cares of the everyday world. Augustine likens "chaste fear" to the feelings of a chaste wife, who rather than dreading that her husband *will return*, anxiously awaits her husband with an apprehension that he will *not* return. This "chaste fear" or passional Angst is interpreted by Augustine as an apprehension of losing the Christian relationship once it is established. Says Augustine: "And as His tarrying is now feared, so after this coming, His leaving will be feared. That will be chaste fear, for it is tranquil and secure."/59/ We may conclude our discussion of Augustine with a brief observation that will prove valuable in the the next chapter, namely, for Augustine man's heart is an abyss, an interpretation that was to deeply touch Eckhart and his followers. Augustine says:

> If by "abyss" we understand a great depth, is not a man's heart an abyss. For what is there more profound than that abyss? Men may speak, may be seen by the operation of their members, may be heard speaking: but whose heart is seen into? What he is inwardly engaged on, what he is inwardly capable of, what he is inwardly doing, or what proposing, what he is inwardly wishing to happen or not to happen, who shall comprehend?/60/

The abyss metaphor is thus for Augustine related to the inwardness of man's soul, to his freedom to choose his own concerns, and to his openness to the possibilities before him. While Augustine's notion of the abyss does not appear directly related to passional Angst, it is worth mentioning because it sets the stage for many later interpretations of Angst and the abyss, especially as regards Kierkegaard's notion of Angst as dizziness. "One may liken dread to dizziness" says Kierkegaard. "He whose eye chances to look down into the yawning abyss becomes dizzy. . . . Thus dread [Angst] is the dizziness of freedom."/61/

Meister Eckhart

Eckhart's pantheistic mysticism has influenced not only German metaphysics but Heideggerian phenomenological ontology in particular. Thus, in order to understand Heidegger's Angst interpretation we must first grasp the role Angst plays in Eckhart's thought. We believe Angst may be the bedrock in Eckhart's thinking, but it becomes visible only through phenomenological analyses of ultimate detachment from the world, the will, and from God himself.

Eckhart's thought is not an elaborate metaphysical system. Rather our knowledge of Eckhart's mysticism is from accounts by Dominican nuns who attempted to transcribe his sermons as Eckhart preached to them. We may find it useful, therefore, to examine three general features of his thought to better grasp Eckhart's unique form of passional Angst. These features are: (1) the distinction between the Godhead and God, (2) the notion of the soul and its two parts, and (3) the description of how the soul unites with the Deity in utter detachment.

First, concerning the Godhead and the Christian God, Eckhart held that a basic distinction must be drawn between God and the Godhead. The God of the Church, the Lord of Creation to whom one prays, is a *living* God. He is the Holy Trinity, namely, God the Father (as the Supreme Deity), God the Son (as the *Logos* or "Word" or uttered thought of the thinker), and God the Holy Spirit (as the love between the Father and the Son). Knowledge of God is possible in two ways: (1) cataphantically, by means of the *via negativa*, which results in a limited knowledge of God at best, or (2) apophantically, by means of a divine union with the Godhead, one achieved only in pure detachment, the total extinguishing of desire and self-will. This results in mystical union or pure knowledge of God.

The Godhead, on the other hand, is well beyond the range of man's discursive intellect. The Godhead is an "unnatured nature," an unplumbed "abyss" (*Abgrund*)/62/ that is at once the totality of all Being and Nothingness. As totality the *Abgrund* is Being in general. All creatures (being) *have* Being—that is, they participate in the Being of the Godhead or the Divine Abyss. But, says Eckhart, the Godhead and God ". . . are as different as heaven and earth."/63/ The Godhead is a self-revealing process which takes place in a single moment—the Eternal Now. Such a process is an outpouring, an emanation occuring when the Word, the *Logos*, is uttered. The Godhead utters, the word "God" is uttered, thus establishing a separation between the two. The "uttered" God becomes the divine Subject: a triune essence of the Father, the Son, and the Holy Spirit. But the Godhead who utters is completely transmundane, so the discursive intellect cannot reach Him. Thus, knowledge of the Godhead is of a different order than knowledge of God. Knowledge of the Godhead must be apophantically given, what Eckhart called an "unknowing knowledge" revealed in Christ as "Divine Knowledge."/64/

The essence of *Divinitas*, or the God within the Godhead, is divine detachment (*Abgeschiedenheit*). Says Eckhart: "For God is God because of His immovable detachment; and from detachment

He has His purity, His simplicity and His unchangeability."/65/ Detachment is likewise a property of Being itself conceived as *divinitas*.

Third, regarding how the soul unites with the Deity, man lives in two realms of Being, says Eckhart: the "inner man" and the "outer man."/66/ The outer man is concerned chiefly with the phenomenal world of the senses: he is the busy man, the man of ontic hustle and bustle, the man who burns up the soul's power in pursuit of sensual experience. The inner man, on the other hand, uses his inner powers to guide his worldly concerns as is necessary, but has the strength to detach himself completely from the world to engage in mystical contemplation of the Godhead. Such detachment takes the form of a self-renouncing abandonment of worldly concerns attained in a destruction of self-will.

Thus mystic union is totally contingent upon renouncing self-will in the form of detachment. Union can occur only when the soul has readied itself. Eckhart describes this readiness as follows:

> Now our Lord says: "no-one hears My word or My doctrine unless he has abandoned himself. For he who would hear the Word of God must be completely self-abandoned [detached]." In the eternal Word that which hears is the same as that which is heard. All that the eternal Father teaches is His being and His nature and His Godhead. He reveals this fully to us in His only begotten Son, and He teaches us that we are the Son.
>
> If a man had emptied himself completely, in such a way that he had become the only-begotten Son, he would possess what is possessed by the only begotten son. . . . When God sees that we are the only begotten Son, he hastens toward us so eagerly, and acts as if His Divine Being would break asunder and be annihilated in itself, in order that He may reveal to us the whole *abyss of His Godhead* and the fullness of His being and of His Nature. Then God hastens toward us so that it may be our own, just as it is God's own. In this God has joy and happiness in abundance. Such a man dwells in the knowledge of God and the love of God, and becomes none other than what God himself is./67/ (Italics added.)

The role of passional Angst in this passage is, we hope, obvious. Detachment means the abandonment of self-will so the soul can emerge its will with God's will and thereby attain union with the Godhead. This, only in the pre-reflective apprehension of the limits to self-will can detachment become possible. Eckhart says: ". . . as long as the craving for more and more is in you, God can never dwell nor work in you."/68/

From a hermeneutical standpoint, Eckhart's thought shows passional Angst as a pre-condition for a new transcendence in Dasein; namely, that the realization that self-will or craving for union with the Godhead is wholly counterproductive. Indeed, self-will must be abandoned in favor of detachment and the humility necessary for Grace. Such transcendence is not possible save through passional Angst, the primordial pre-disposition through which mystical Dasein is born again. The Grace of the Godhead, says Eckhart, is given only to the truly humble: It can never be the product of self-will. Thus, through passional Angst, mystical Dasein is united in the No-thingness of the Divine Abyss: its divine spark is merged with the self-same essence of the Godhead. Mystical union is attained. We will return to a discussion of this in our final chapter.

In Eckhart, then, passional Angst shows itself in a striking and powerful culmination of the Augustinian tradition. But there remains one final figure who, technically speaking, stands outside this tradition but is one who has a profound sensitivity to passional Angst. Thus, to this final figure, Jacob Boehme, we now turn, for his contributions to mystical thought influence the post-Kantian German Idealists, who in turn influenced Søren Kierkegaard.

Jacob Boehme

The final figure to be considered in this section is Jacob Boehme (1575–1624). Boehme's contribution to Western philosophy has largely been either overlooked or covered over due perhaps to his obscurity as well as to his pantheistic and alchemical nature mysticism. Yet, Boehme's essential insights into the ultimate Yes and No of Being, his formulation of the great dialectic, as Hegel understood the term, assures him an important place in the history of Western thought. William Inge notes, for example, that Sir Isaac Newton "shut himself up for three months to study Boehme, whose teaching on attraction and the laws of motion seemed to him to have great value."/69/ Too, Boehme's influence on the post-Kantian Idealists such as Franz von Baader, Schelling, and Hegel was profound.

Jacob Boehme was born, raised, and buried a Lutheran, and even in his most alchemical period he still considered himself as living within the framework of Lutheran doctrine./70/ Thus to fully appreciate the scope of Boehme's thought, we see it within the joint context of (1) Theophrastus Bombastus Paracelsus (1493–1541), Sebastian Franck (1500–1545), and Valentine Weigel (1553–1588) as advocates of Nature Mysticism; and of (2) Luther's doctrines of personal piety, the opposition of love and wrath in God, the problem of justification, freedom of the will, and textual interpretation of creation stories./71/

To see the role of passional Angst and the abyss in Boehme's thought is not at all difficult. Indeed, it is on the very surface of his doctrine of the Fall of Man. Yet to properly understand passional Angst and the abyss in Boehme's writings, we must begin with a cursory outline of three important dimensions of his thought. These are: (1) the doctrine of the Godhead as the *Ungrund* or abyss, (2) the doctrine of *Urwille* or the free will of the Godhead that gives rise to generation and the dialectic, and (3) the doctrine of the Fall of Man and its consequences for the soul.

First, Boehme's doctrine of the Godhead is a direct result of his profound mystical experience, which came during the spring of 1600./72/ In his Epistles Boehme describes his first mystical encounter as follows:

> For I saw and knew the Being of all beings, the ground and the unground (*Ungrund*); the birth of the holy trinity; the source and origin of this world and all creatures in divine Wisdom (Sophia). . . . I saw all three worlds in myself, (1) the divine, angelical, or paradisaical; . . . (2) the dark world . . . ; (3) the external visible world . . . ; and I saw and knew the whole Being in evil and good, how one originates in the other . . . so that I not only wondered but also greatly rejoiced./73/

In his vision the whole of the Godhead as the Divine *Ungrund* or abyss was revealed to Boehme. Thus, he believed the *Ungrund* was there from eternity, a pure no-thingness (*Nichts*) dwelling in the absolute silence of eternity: self-contained, a complete unity without distinctions, unknowing and unknowable, the great hidden and invisible./74/ Meontic, the Godhead is beyond all dichotomies of good and evil, Yes and

No, freedom and desires. Yet all dichotomies are contained within the Godhead.

But, claimed Boehme, in the core of the Godhead, where all dichotomies are united, there resides a dark and irrational craving for self-revelation, for nature as creation, for a ground to the *Ungrund*. We are told that this doctrine was derived from Boehme's interpretation of the gospel according to John./75/

> The nothing hungers after the something, and his hunger is the desire. . . . For the desire has nothing that is able to conceive. It conceives only itself, and draws itself to itself . . . and brings itself from abyss to byss (*vom Ungrunde in Grund*) . . . and yet remains a nothing./76/

Thus the craving or desire for the *Ungrund*'s self-knowledge and self-subjectification and its opposite but equal craving for self-emanation and objectification; all this is the first instance of the great dialectic of Being. This great dialectic is instigated by the dark and irrational *Urwille* or primordial will of the abyss. From this initial step, the *Ungrund* or abyss of the Godhead manifests itself as God in a primal act of self-consciousness.

The doctrine of the *Urwille*, our second point, is in itself dialectical. On the one hand the *Urwille* is the free, untainted essence of the abyss or *Ungrund*. But on the other hand, the *Urwille* is a form of the Godhead's craving for self-knowledge as it manifests itself as nature. This is the second instance of the great dialectic, the Yes and the No of all Being:

> The One, as the Yes, is pure power and life, and is the truth of God, or God himself. He would in Himself be unknowable, and in Him would be no joy of elevation, nor feeling, without the No. The No is the counterstroke of the Yes, or the Truth, in order that the Truth may be manifest and a something, in which there may be a contrarium, in which the eternal love may be loving, feeling, willing, and such as can be loved./77/

Thus, the *Urwille* sets into motion the dialectical forces of the great Yes and No of Being: As free-will in the Godhead, the *Urwille* is manifested as love and grace; as craving or desire in the Godhead it is manifested as wrath and darkness.

Concerning our third topic, namely, the doctrine of the Fall of man and its consequences for the soul, we may begin with

Boehme's description of the "essential man" (*Urmensch*). God made man as a being who suffered no strife, no earthliness, and no flesh, says Boehme. *Urmensch* did have a body, but it was pure and completely transparent to the celestial light./78/ Such a being was androgynous, that is, a mixture of the inner and outer worlds, was lord over creation, and was possessed with free-will. This last feature, *Urmensch*'s free-will, made possible the Fall of man. Specifically the Fall took place, says Boehme, in Adam's sleep. According to Boehme's understanding, in Adam's pure state of Being before the Fall, he did not sleep, since the dialectic of sleeping and wakefulness had not yet risen. Thus, Adam gazed with open eyes at the glory of divine consciousness. But as a creature of God he shared the *Ungrund*, so stirring in his free will was a craving for disunity, for self-knowledge, and for the knowledge of good and evil./79/ From his standpoint in eternity, Adam turned away from union with God, slipped into sleep, and awoke in the temporal realm (Time). As he awoke he found he had emerged into a new form of being, says Boehme; namely an Angst-filled existence. Here for the first time in history, ontological Angst is named.

What does Angst mean for Boehme? It is discussed in five of his writings in various guises: In *Aurora* (xiii, 118) it means the cause of sadness and joy in man; in *Von der Menschwerdung Jesu Christi* (II, iii. 13), Angst is what prevents eternal darkness; in *Zweite Schutzschrift gegen Balthasar Tilke* (141), Angst is an idle nothing; in *Von der Gnadenwahl* (111, 5), Angst is man's longing for primordial freedom as he lives a life of torment and death; and then in *Tafel der Drei Principien* (39), Angst is the well-spring of hell-fire in the soul of a sensitive man./80/ But the most complete description of Angst is provided in *Von der Gnadenwahl* where Boehme says Angst is:

> A root of feeling, the beginning . . . of mind, a root of . . . all painfulness . . . a manifestation of the eternal unfathomable will in the attraction . . . a cause of dying . . . the very root where God and Nature are separated . . . [where] the manifest sensible eye arises./81/

The reference here to "a root of feeling, the beginning of mind" is extremely provocative. While it may be tempting to think that what Boehme is describing here is a pre-disposition or attunement (*Befindlichkeit*) which provides access to being-in-the-world, such hermeneutical interpretations cannot be sound, for Boehme's Angst is not yet the phenomenological Angst of twentieth-century thought. Rather it is a vivid depiction of the passional Angst of mysticism.

Boehme's conception of Angst is related to the Fall of man, resulting from man's free-will. According to Boehme, man (*Urmensch*) fell due to a sort of cosmic or angelic boredom with the unitive life. Adam as essential man slipped into the existence from a craving for the knowledge of Good and Evil. He chose, and in so choosing became Dasein, a being-in-the-world. Boehme's passional Angst, then, is the apprehension of the loss of God's grace due to a deliberate act of self-will. Passional Angst reveals in Boehme's thought a barrier between the soul and the Godhead, resulting from the soul's willfulness and ontological craving.

In Boehme passional Angst is clearly seen as a facet of primordial Angst. For Boehme the "longing for primordial freedom" in the abyss of the Godhead is heard as a plaintive cry for a return to unity in the Godhead. Boehme says, therefore, Angst is: ". . . the tension between man's two wills. Man remembers his lost unity; he longs for his primordial freedom as he lives a life of torment and death."/82/

This powerful and compelling interpretation of Angst ultimately became an indirect source for Kierkegaard's *Begrebnet Angst* through the writings of Schelling, von Baader, and principally Hamann. Thus, in the next chapter, we shall discuss Kierkegaard's concept of Angst in the light of these nineteenth century interpreters of Boehme and the passional Angst of the mystical tradition of western thought.

NOTES

/1/ Jacob Boehme, *Aurora oder Die Morgenröthe im Aufgang*, xiii, 118, as cited in John J. Stoudt, *Jacob Boehme: His Life and His Thought* (New York: The Seabury Press, 1968).

/2/ Søren Kierkegaard, *Begrebnet Angest*, translated by Walter Lowrie as *The Concept of Dread* (Princeton: Princeton University Press, 1944), and by Reidar Thomte as *The Concept of Anxiety* (Princeton: Princeton University Press, 1980).

/3/ "Indo-European Roots," *The American Heritage Dictionary*, New College Edition, s.v. "*angh-*."

/4/ Ibid.

/5/ James Hillman, "On the Necessity of Abnormal Psychology," *Eranos* 43 (Leiden: Brill, 1947), p. 97 (hereafter cited as ONAP).

/6/ Heinz Schreckenberg, *Ananke—Untersuchungen zur Geschichte des Wortgebrauchs, Zetemata*, Heft 36 (Munchen: Beck, 1964), pp. 169–74.

/7/ Hillman, ONAP, pp. 97–98; cf. the "Special Lexicon" inclusion where *Anchein* is translated as "to squeeze."

/8/ Henri Frankfort et al., *Before Philosophy* (Baltimore: Penguin Books, 1964), pp. 23–36.

/9/ Ibid., p. 14.

/10/ Cornelius Loew, *Myth, Sacred History and Philosophy: The Pre-Christian Religious Heritage of the West* (New York: Harcourt, Brace and World, Inc., 1967), p. 20.

/11/ Ibid., p. 19. Note: cf. Frankfort et al., *Before Philosophy*, pp. 155–57.

/12/ Indeed, such an account reveals a great deal about the role of Dasein in the Mesopotamian Cosmos. For such Dasein the Cosmos was conceived as a vast universal political State. Dasein's proper role was obedience at every level, according to Frankfort. Thus, the presuppositions of the time colored the understanding, disposition, and articulation of Ancient Mesopotamian Dasein much as does the technological presuppositions of our own time, namely, that we are captives of our own presuppositions about the nature of Being. Cf. Frankfort et al., *Before Philosophy*, pp. 217–33.

/13/ Frankfort et al., *Before Philosophy*, p. 59. Note: to be sure, this is not the only creation myth of Egypt. The Memphite theology tells an analogous tale with emphasis in the role of "the Word" in the creation process. Too , there are other creation myths such as that offered in the *Book of the Dead*. The one discussed here is therefore representative; cf. Frankfort et al., *Before Philosophy*, p. 60, for other accounts.

/14/ Ibid., p. 63.

/15/ Ibid., p. 61.

/16/ Genesis 1:1, New English Bible (hereafter cited as NEB).

/17/ Gerhard von Rad, *Genesis: A Commentary*, trans. by John H. Marks (Philadelphia: The Westminster Press, 1961), p. 48. It is interesting to note that Frankfort dismisses von Rad's argument as ". . . interesting but not too alluring because the Egyptian story and the Hebrew diverge immediately when one comes to the episodes of creation, with Egypt emphasizing the self-emergence of a creator-God, whereas the creator-god of Genesis existed alongside the chaos"; cf. Frankfort et al., *Before Philosophy*, p. 61.

/18/ Frankfort et al., *Before Philosophy*, p. 249.

/19/ F. M. Cornford, *From Religion to Philosphy: A Study in the Origins of Western Speculation* (New York: Harper & Row, 1957), p. 17.

/20/ Plato, *Timaeus 30a 4–5*. Translated by B. Joweth. *The Works of Plato* (New York: The Dial Press, 1936).

/21/ Frankfort et al., *Before Philosophy*, p. 220.

/22/ Frankfort admits that there is a possibility that he has mistranslated Ma'at as "justice," "truth," or "righteousness." It is equally possible, he notes, for Ma 'at to mean "order," "regularly," or "conformity." His choice of the former set, as he again freely admits, is based upon a subjective judgment grounded in his general understanding of Egyptian culture. Thus free admission is, of course, consonant with sound hermeneutical practices. See Frankfort et al., *Before Philosophy*, p. 116.

/23/ Gerhard von Rad, *Wisdom in Israel* (Nashville and New York: Abington Press, 1972), p. 72.

/24/ Ibid.

/25/ Thomas Katsaros and Nathaniel Kaplan, *The Western Mystical Tradition: An Intellectual History of Western Civilization*, vol. I (New Haven: College and University Press, 1969), p. 33, note 1 (hereafter cited as *WMT*).

/26/ Cornford, *From Religion to Philosophy*, p. 127.

/27/ Ibid., p. 128. Cornford points out that Aristotle's interest in the first proposition of Thales overshadowed the second and third propositions. Here is a clear example of covering-over fundamental (indeed, in this case *founding*) concepts of philosophy.

/28/ W. G. Lambert, *Babylonian Wisdom Literature* (Oxford: Oxford University Press, 1960), p. 10, as cited in Scott, *The Way of Wisdom*, p. 143.

/29/ Samuel N. Kramer, "Man and His God" (1955), pp. 153–55, as cited in Scott, *The Way of Wisdom*, p. 143.

/30/ *Ancient Near Eastern Texts Relating to the Old Testament*, edited by J. B. Prichard, 2d ed. (Princeton University Press, 1955), pp. 434–37 (hereafter abbreviated as *ANET*).

/31/ Ibid.

/32/ Ibid., pp. 438–40.

/33/ R. B. Y. Scott, *The Way of Wisdom in the Old Testament*, edited by J. B. Pritchard, 2d ed. (Princeton: Princeton University Press, 1955), p. 43.

/34/ Ibid.

/35/ *ANET*, pp. 405–7.

/36/ Ibid.

/37/ Job 30:15–21. (NEB).

/38/ F. M. Cornford, *Plato's Cosmology The Timaeus of Plato* (London: Routledge, 1948), p. 160, as cited in Hillman, ONAP, p. 108.

/39/ Hillman, ONAP, p. 108.

/40/ Ibid., p. 98.

/41/ Ibid., p. 99.

/42/ R. B. Owens, "The Knees of the Gods," *The Origins of European Thought* (Cambridge: Cambridge University Press, 1954), pp. 303–9; cf. Hillman, ONAP, p. 111.

/43/ Aristotle, *Metaphysics, The Works of Aristotle*, 2d ed., translated by W. D. Ross (Oxford: Clarendon, 1940), 1015a; cf. Hillman, ONAP, p. 106.

/44/ Hillman, ONAP, p. 106.

/45/ Aeschylus, "Prometheus Bound," as cited in Hillman, ONAP, pp. 100–101.

/46/ Euripides, "Alcestis," as cited in Hillman, ONAP, p. 102.

/47/ Hans Jonas, *The Gnostic Religion*, 2d ed. (Boston: Beacon Press, 1963), p. 25. Note: we are indebted to this work for much of the discussion of Gnosticism. Professor Jonas was one of Heidegger's students and is a hermeneutic phenomenologist par excellence in his own right. Jonas himself has not failed to see the Angst of gnostic thought and explicitly recognizes the prima facie connection between Heidegger's thought and Gnosticism, but cautions against concluding from this that the former is derived from the latter.

/48/ Jonas, *The Gnostic Religion*, pp. 65ff.

/49/ Susan A. Taubes, "The Gnostic Foundations of Heidegger's Nihilism," *The Journal of Religion* 34 (July 1954): 158–59.

/50/ Jonas, *The Gnostic Religion*, p. 49.

/51/ Robert M. Grant, *Gnosticism: A Source Book of Heretical Writings From the Early Christian Period* (New York: Harper and Brothers, 1961), p. 115.

/52/ *The Nag Hammadi Library*, ed. James M. Robinson, translated by the members of the Coptic Gnostic Library Project (San Francisco: Harper & Row, 1977), p. 74.

/53/ Jonas, *The Gnostic Religion*, pp. 320ff.

/54/ Taubes, "The Gnostic Foundations of Heidegger's Nihilism," pp. 155ff.

/55/ This was, of course, not surprising. Gnosticism was the great competitor to apostolic Christianity and at one point had a far greater number of converts than did the early Christian church. Willison Walker, for example, believes the "Gnostic Crisis" (i.e., the challenge of gnostic thought) was responsible for the development of the Apostolic Creed, as well as the coalescence of power in the Bishops of Rome, Smyrna, or Ephesus. Thus, upon the challenge of Gnosticism is the Catholic Church built, according to Walker; cf. Willison Walker, *The History of the Christian Church*, rev. ed. (New York: Scribners, 1958), pp. 58–59.

/56/ Augustine, *The Confessions of St. Augustine*, translated by E. B. Pussey (New York: Airmont Publishing Company, 1969), Book VIII, xii, 28.

/57/ Ibid.

/58/ Augustine, *Enarrationes in Psalmos* XXVII, 80, as cited in Erich Przywara, *An Augustine Synthesis* (Gloucester, MA: Peter Smith, 1970), pp. 415–16.

/59/ Ibid.

/60/ Ibid., XLI13, as cited in Przywara, *An Augustine Synthesis*, p. 421.

/61/ Kierkegaard, *CD*, p. 55.

/62/ Meister Eckhart, "Sermon XX," as cited in James M. Clark, *Meister Eckhart: An Introduction to the Study of His Works with an Anthology of His Sermons* (London: Thomas Nelson and Sons, 1957), p. 224. It is this notion of the difference between the God and the Godhead that led to Eckhart's condemnation as a pantheist. There is no doubt that his pantheistic mysticism differs markedly from the Christ-mysticism of St. John of the Cross. Likewise, therefore, primordial Angst becomes schematized very differently in Eckhart than in John of the Cross. In Eckhart, passional Angst is more highminded and even intellectual pantheism, whereas in John of the Cross it appears to be passionate in all senses of that term, but specifically with regard to Christ's passion. The latter's mysticism is the Christ mysticism of a theistic verus pantheistic conception of Being.

/63/ Meister Eckhart, "Sermon XII," as cited in James M. Clark, *Meister Eckhart*, p. 41.

/64/ C. F. Kelley, *Meister Eckhart on Divine Knowledge* (New Haven and London: Yale University Press, 1977), p. 7. Note: this work is a highly technical analysis of the concept of Divine Knowledge in Eckhart. It clears up much of the conceptual confusion that surrounds Eckhart's thought that results more from Eckhart's mystical language than from the difficulty of his thought.

/65/ *Meister Eckhart: Selected Treaties and Sermons*, translated by J. M. Clark and J. V. Skinner (London: Farber and Farber, 1958), p. 164, as cited in John D. Caputo, *The Mystical Element in Heidegger's Thought* (Drexel Hill, Pennsylvania: Oberlin Publishing Co., 1978), p. 12 (hereafter cited as *MEHT*).

/66/ Franz P. Pfeiffer, *Meister Eckhart: Selected Treatises and Sermons*, translated by J. M. Clark and J. V. Skinner (London: Farber and Farber, 1958).

/67/ Caputo, *MEHT*, pp. 14–15. We are much indebted to Caputo's insights in this section, especially on his interpretation of the soul, detachment, and the Godhead in Eckhart's thought.

/68/ Clark, *Meister Eckhart*, Sermon XXX, pp. 223–24.

/69/ W. R. Inge, *Christian Mysticism* (London: Methuen and Co., 1918), p. 278.

/70/ Stoudt, *Jacob Boehme*, p. 147.

/71/ Ibid., p. 149.

/72/ Ibid., p. 56.

/73/ Jacob Boehme, *Epistles*, xii, 8, as cited in Stoudt, *Jacob Boehme*, p. 198.

/74/ Boehme, *Mysterium Magnum*, preface, 6, as cited in Stoudt, *Jacob Boehme*, p. 200.

/75/ Specifically John 1:1–3. Cf. Boehme, *Von der Gnadenwahl*, ii, 7–11, as cited in Stoudt, *Jacob Boehme*, p. 198.

/76/ Boehme, *Mysterium Magnum*, iii, 5, as cited in Stoudt, *Jacob Boehme*, p. 200.

/77/ Boehme, *Von 177 Theosophischen Fragen*, iii, 2, as cited in Stoudt, *Jacob Boehme*, p. 205.

/78/ Boehme, *Von der Geburt und Bezeichnung aller Wesen*, xi, 51, as cited in Stoudt, *Jacob Boehme*, p. 205.

/79/ Boehme, *Mysterium Magnum*, xviii, 28, as cited in Stoudt, *Jacob Boehme*, p. 266.

/80/ Stoudt, *Jacob Boehme*, p. 270.

/81/ Boehme, *Von der Gnadenwahl*, iii, 5, as cited in Stoudt, *Jacob Boehme*, p. 270.

/82/ Ibid.

CHAPTER II

KIERKEGAARD AND ANGST

> Deep within every human being there still lives the anxiety [Angst] over the possibility of being alone in the world, forgotten by God, overlooked among the millions and millions in his enormous household.
>
> —Søren Kierkegaard, 1837

The Question at Hand

On a storm-drenched afternoon in 1767 somewhere on the wind swept heaths of Jutland, an eleven-year-old shepherd boy sits huddled at the top of a hillock, grimly watching his sheep on the heath below him. A flash of lightning reveals his pinched and angry features: his wind tousled hair, his twisted lips, his red-rimmed eyes lifting upwards to the lead grey sky. Slowly he rises, his small hands clenching into fists, a rage growing within him. Another lightning bolt. He screams, shaking his white-knuckled fist at the heavens. The booming thunder obliterates his impassioned curse against the Holy Ghost for allowing the hopelessness, the poverty, the hunger, and the loneliness into his young life.

Here, in a moment of profound rage, the boy who became Søren Kierkegaard's father first felt the stirrings of primordial Angst. For Michael Kierkegaard, however, the experience was the unforgivable sin of cursing the Holy Ghost./1/ His abysmal guilt, his utter self-loathing for committing the sin that must pass on to his children and their children, made this Jutland Dane a morbid, melancholic man imbued with an inordinant sense of guilt. Michael Kierkegaard sincerely believed that he and all that was of his flesh were destined for eternal damnation. This became the "dreadful secret" he was incapable of revealing to anyone for decades of his life.

Only in 1844, over seventy-five years after the Jutland experience, was it possible for Søren Kierkegaard to understand the condition which made his father's sin possible. Indeed, it was not his father's general anger at the vicissitudes of the shepherd's life, nor was it even the frustration he felt at the poverty, isolation, and hopelessness of his life on the Jutland heath. Rather, the possibility of cursing the Holy Ghost came from within his father's spirit as a manifestation of what we have called primordial Angst—the Angst in the face of an abyss of meaninglessness.

Our purpose in this chapter, therefore, is to show how Søren Kierkegaard came to understand Angst as the ground of not only his father's sin, but of sin in general. To accomplish this purpose, however, several guidelines for interpretation must be mentioned so as not to mislead the reader; for in entering upon a discussion of Kierkegaard's concept of Angst, we depart from the previous chapter's macro-hermeneutical interpretations and enter into a micro-hermeneutical approach to Angst.

Thus in the pages to follow, we shall set the stage for a micro-hermeneutical as opposed to a macro-hermeneutical approach to Angst. We will begin with an examination of the overall background of Kierkegaard's method of philosophical investigation. Only when this task has been worked through, we suggest, can the "figure" of the phenomenon of Angst show itself against this "background," somewhat as would a figure appear against the ground in terms of Gestalt theory.

We begin our micro-hermeneutical task by turning to the first in-depth treatment of Angst as a theme for philosophical investigation. The work to which we refer was the product of Virgilius Haufniensis, a Kierkegaardian pseudonymn meaning significantly enough, "watchman of Copenhagen." This seminal work was entitled *Begrebnet Angest* [*The Concept of Angst*], subtitled *A Simple Psychologically Orienting Deliberation on the Dogmatic Issue of Hereditary Sin.*

Often called the "father of Existentialism,"/2/ Kierkegaard's impact on contemporary thought—particularly contemporary European thought—can hardly be overestimated. Certainly twentieth-century phenomenological ontologists and existential thinkers, whether they agree with Kierkegaard or not, owe a profound debt to his analysis and description of Angst and human freedom. As Rollo May has shown,/3/ an entire school of existential psychiatry is deeply grounded in Kierkegaard's work on Angst and specifically its relationship to the demonic.

In view of Kierkegaard's considerable influence, especially

with respect to Angst, it is truly tempting to seek to interpret the whole of his pseudonymous writings in the light of Angst. For Kierkegaard, Angst is the psychological pre-condition for human freedom. It is man's departure point, his telos: mental-spiritual development toward Christianity and the eternal. But if this general interpretation is tempting, it is also misleading; for *any* attempt to distill the essence of Kierkegaard's aesthetic writings to a single concept is nothing short of a flagrant reductionism. As Howard V. and Edna H. Hong observed, Kierkegaard must be read "organically and collaterally, not linearly or atomistically as some hapless writers on Kierkegaard have suggested."/4/ The almost uniform perspective of such hapless writers, the Hongs continue, is an appeal to ". . . a single work (usually a pseudonymous work) as a means for interpretation and critique."

With this observation in mind, we decline to join the ranks of those hapless writers. Rather, in this chapter we hope to show that in Kierkegaard's work Angst must be approached collaterally and organically rather than linearly or atomistically. Thus, we will here approach Angst from three distinct but related points of reference: (1) from Kierkegaard's general existential method of philosophizing; (2) from the *Journals and Papers* as a reflection of his immediate apprehension of Angst's meaning and significance; and finally (3) from *Begrebnet Angest* itself. This should provide a collateral hermeneutical context for understanding Kierkegaard's aesthetic or pseudonymous treatment of the Angst phenomenon.

Our micro-hermeneutical horizons may be broadened if one preliminary issue is discussed before launching into our major tasks. This is the question of the *direct influences* on Kierkegaard as the basis for his Angst interpretation. Not only is this question interesting in itself, but it speaks to the significant question of Kierkegaard's *originality* regarding Angst.

The Horizons of Influences

The question of the influences on Kierkegaard's analysis of Angst is difficult to unravel for several reasons. First, from Kierkegaard's *Journal* entries and the footnotes within *Begrebnet Angest* we find he was familiar with the basic writings of the Christian Church Fathers,/5/ as well as with the gnostic tradition./6/ Likewise Kierkegaard was acquainted with some of

Jacob Boehme's thought/7/ and refers to St. John of the Cross on one occasion./8/ Furthermore, a good possibility exists that Kierkegaard became acquainted with Meister Eckhart's thought through Franz von Baader's works, particularly his *Vorlesungen . . . ueber religiose Philosophie* [*Lectures Concerning Religious Philosophy*]/9/ although this is not confirmed in the *Journals and Papers*./10/ Be that as it may, Kierkegaard, as an intellectual product of nineteenth-century thought, was certainly well acquainted with the works of his immediate predecessors, J. G. Fichte, F. W. J. Schelling, F. E. D. Schleiermacher, J. G. Hamann, and of course G. W. F. Hegel, all of whom were capable, at least in principle, of directly influencing Kierkegaard with regard to the interpretation of *Begrebnet Angest*.

Concerning particular textual reference to such influences, Kierkegaard indirectly cites von Baader in *Begrebnet Angest* and directly discusses the contributions of Schelling and Hamann to the development of his concept of Angst. But of all these thinkers, Hamann appears to have influenced Kierkegaard the most. Specifically, as early as 1842 Kierkegaard wrote in his *Journals*:

> In Volume VI, p. 194, of his works, Hamann makes an observation which I can use, although he neither understood it as I wish to understand it nor thought further about it: "However, this Angst in the world is only proof of our heterogeneity. If we lacked nothing, we should do no better than the pagans and the transcendental philosophers, who know nothing of God and like fools fall in love with lovely nature, and no homesickness would overcome us. This impertinent disquiet, this holy hypochondria. . . ."/11/

The passage is concluded in the last footnote to *Begrebnet Angest*:

> ". . . is perhaps the fire with which we season sacrificial animals in order to preserve us from the putrefaction of current *secunda*."/12/

While Hamann did influence Kierkegaard concerning the meaning of Angst, he did so only *indirectly*, or perhaps more accurately, *collaterally*; for while Hamann's comments may have been useful to Kierkegaard, Kierkegaard did not understand the phenomenon of Angst in the same way as did Hamann.

This gives rise to an even deeper question: how did Kierkegaard "wish to understand" the phenomenon of Angst by 1842? The answer to this question seems clear. The *Journal* entry cited above is an addendum to an earlier entry (likely written in sequence), in which Kierkegaard describes for the first time the significance of Angst in relation to original sin:

> The nature of original sin has often been explained, and still a primary category is lacking—this is anxiety [Angst]; this is the essential determinant. *Anxiety is a desire for what one fears*, a sympathetic antipathy; anxiety is an alien power which grips the individual, and yet one cannot tear himself free from it and does not want to, for one fears, but what he fears he desires. Anxiety makes the individual powerless, and the first sin always occurs in weakness; therefore it apparently lacks accountability, but this lack is the real trap./13/ (Italics added.)

Thus as early as 1842, Kierkegaard understood Angst's ambiguous power, an understanding that was worked out concretely in *Begrebnet Angest* two years later. Our only point here is to show Kierkegaard's Angst interpretation as "sympathetic antipathy" was markedly different from Hamann's "holy hypochondria" Angst.

Another contributor to the literature on Angst was Schelling. In an important footnote within *Begrebnet Angest*, the pseudonymous author Virgilius Haufniensis observes:

> Schelling himself has often spoken of anxiety [Angst] anger, anguish, suffering, etc. But one ought always to be a little suspicious of such expressions, so as not to confuse the consequences of sin in creation with what Schelling also characterizes as states and moods in God. . . . Schelling's main thought is that anxiety etc., characterize especially the suffering of the deity in his endeavor to create. (*CA*, p. 59, note)

Having said that, Haufniensis virtually dismisses Schelling's Angst contribution. More interesting perhaps is the unpublished draft of the above passage where Jacob Boehme is linked with Schelling. The draft notes in the margin: "Jacob Boehme, Schelling. 'Anxiety [Angst], anger, hunger, suffering.' These things should always be eyed with caution; now it is the consequence of sin, now the negative in God—[the other]."/14/ Therefore, it is clear that both

Schelling and Boehme were *considered by* Kierkegaard in his collateral development of the concept of Angst, but neither were directly responsible for Kierkegaard's Angst interpretation.

One final predecessor, Franz von Baader, demands discussion. While Kierkegaard does not refer to von Baader's concept of Angst directly in either *Begrebnet Angest* or in the *Journals and Papers*, he does criticize von Baader on two principal points in footnotes to *Begrebnet Angest*. In one footnote, for example, Haufniensis says ". . . concerning the significance of temptation for the consolation of freedom," von Baader overlooks an important immediate term in the transition from innocence to guilt. Significantly enough, this immediate term is Angst. In a second footnote, Kierkegaard criticizes von Baader as a thinker who ". . . did not take into account the history of the race," a crucial point in the discussion of "objective Angst" in *Begrebnet Angest*. From Haufniensis' viewpoint, von Baader's omission (in tandem with his interpretation of finitude and sensuousness as sinfulness), leads von Baader close to the edge of the Pelagianistic heresy,/15/ a position to which Haufniensis (and Kierkegaard himself) was unquestionably opposed.

In view of this evidence, it may be more fruitful to start from the premise that Kierkegaard's Angst concept was not the product of any direct influence. Rather, we believe that the Angst of *Begrebnet Angest* first occurred to Kierkegaard as a personal experience rather than as a cognitive idea. Only after many years of struggling with, indeed, *consistently thinking through*, the implications of Angst in this personal life, was Kierkegaard able to formulate the concept that becomes the essential theme of *Begrebnet Angest*

This interpretation is, of course, only the skeletal structure which this chapter seeks to flesh out. But to accomplish our aim, we must begin by attempting to grasp Kierkegaard's general method of philosophical investigation; for without this micro-hermeneutical background to the approach to the concept of Angst, *Begrebnet Angest* becomes an enormously difficult work to understand, interpret, and discuss.

As our next question, then, we must inquire into Kierkegaard's philosophical method, which in our view is a synthesis of two dialectically opposing perspectives: on the one hand, we may distinguish "the cognitive approach to method," one grounded in the apodicticity of logic and science, and one carried out via the principle of consistent thinking. On the other hand, there is "the experimental/psychological approach to method," one grounded in experimentation, and one carried out via penetrating psychological observations.

But before turning to these matters, one last preliminary point must be mentioned: we are indeed aware of the significant role Angst plays in *Either/Or*, as the "root energy of the aesthete as he tries to flee from finitude with demonic energy,"/16/ of its role in *Fear and Trembling* as the Angst before the demand of religious faith;/17/ and finally of the significance Angst has for *The Sickness Unto Death* as the essence of melancholy and despair as the spirit moves toward its destiny./18/ Yet a full-dress treatment of these manifestations of primordial Angst in Kierkegaard's thought is beyond the scope of the present work. It is our task here, after all, to discover the *essential* role of Angst in Kierkegaard's thought rather than showing how that role is employed throughout the entirety of his pseudonymous authorship.

The Micro-Hermeneutical Horizons of Kierkegaard's Thought

In this section, our task is to provide a horizon for interpreting Angst in Kierkegaard's thought. We have argued that we must approach Kierkegaard's thought both collaterally and organically. Thus, in the following pages we hope to show that the Haufniensis concept of Angst can best be interpreted in the context of Kierkegaard's *method of philosophical investigation*.

We further argued that Kierkegaard's method is neither linear nor atomistic: it is the product of two dialectically opposing approaches which when unified provide a rigorous means of carrying out scholarly investigations within the tradition of the spiritual sciences [*Geisteswissenschaften*]. Kierkegaard was concerned with the totality of the spiritual sciences, seen from the viewpoint of human existence, as his basic intellectual project, and the collateral means of investigating human existence became his "methodology."

To discuss both approaches to Kierkegaard's methodology, we will divide this section into two subsections. First, we shall explore Kierkegaard's cognitive approach to method. Then we will examine his experimental/psychological approach.

The cognitive approach to Kierkegaard's method. An absolutely essential source for understanding Kierkegaard's thought is his *Papirer* (*Journals and Papers*). Perhaps more than either the pseudonymous aesthetic writings or the Christian discourses written under his own name, the *Journals* provide a personal side to Kierkegaard's thought which compliments and strengthens the other dimensions of his authorship.

As the Danish editors of these *Journals and Papers* have shown, as early as 1836, when he was twenty-three years old, Kierkegaard was already coming to grips with a "system" for methodically exploring what he called his "project." Kierkegaard's project was:

> The collection of material for a characterization of the spirit of the Middle Ages through a general historical study of the age's distinctive features in all areas of the spiritual intellectual life, in literature, art, religion, science, and social conditions, concentrating on a more thorough and concrete study of the reflection of the folk genius of the Middle Ages in poetry, legends, fairy tales, and stories, especially on the personifications of the representative ideas rising out of the medieval folk life's world of consciousness: Don Juan, Faust, the Wandering Jew, and all this in the light of a more abstract Hegelian-philosophic parallel interest in a comprehensive delineation of the stages of intellectual-spiritual development, including "world history" as well as the single individual's "microcosm," by way of defining concepts such as: the classical, the romantic ("dialectical"), the modern, comedy, tragedy, iron, humor, resignation, etc., etc./19/

To say the least, this single sentence is astonishingly ambitious. Nonetheless, the description lacks a single thread of coherence to pull together its elegant tapestry of thought. Gregor Malantschuk suggests the unifying thread is the concept of "anthropological contemplation."/20/ Anthropological contemplation was discovered by Kierkegaard in 1838 as the key to philosophizing in the older classical Greek sense, namely, one which ". . . unreflectively assumed as a beginning—that on the whole there is reality in thought."/21/ From the assumption that thought has reality, Kierkegaard concludes, "but the whole line of thought proceeding from this assumption entered into a genuine anthropological contemplation *which has not yet been undertaken*."/22/ (Italics added.)

Kierkegaard's intent was not only to *undertake* the task of anthropological contemplation, but to *carry it out* to the best of his ability. If this task is to be carried out successfully, Kierkegaard reasoned, it would require at minimum an apodictic point of departure from which all other conclusions could be derived, that is, a methodological Archimedian point./23/

Kierkegaard found his Archimedian point in the Socratic principle: "know thyself!" As early as 1835, he was aware of its apodictic power, noting in the *Journal*, "One must first learn to know himself before knowing anything else."/24/ Nine years later, the author of *Begrebnet Angest* echoes this maxim in a different manner: "And this is the wonder of life, that each man who is mindful of himself knows what no science knows, since he knows who he himself is, and this is the profundity of the Greek saying which too long has been understood in the German as pure self-consciousness, the airiness of idealism" (*CA*, p. 79, note). To assert the apodicticity of the "know thyself" dictum is, of course, one thing; but to offer compelling evidence for it is quite another. Nevertheless, Kierkegaard offers a compelling indirect proof of such apodicticity. He argues: we could never have an understanding of humanity in general, that is, the conditions which make being human possible, unless we begin with self-understanding. More specifically, "Every person must fundamentally be assumed to possess that which is fundamental to being a person. . . ."/25/ If this were not the case (the indirect proof continues), then ". . . at various times fundamentally different people have been produced, and the universal unity of mankind is abolished."/26/ This, however, *could not* be the case, since ". . . every person possesses in himself, when he looks carefully, a more complete expression for everything human than the *summa summarium* of all knowledge that he acquires by learned studies."/27/ Therefore (the indirect proof concludes), ". . . what holds here in a profound psychological decision is *unum noris omnes* [if you know one, you know all]. When the possibility of sin [for example] is shown in one, it is shown in all, and all that is left to the ideal arena of observation are considerations of more or less [sin in the individual]."/28/ Thus the ground of Kierkegaard's general philosophical method may be characterized as the principle of *unum noris omnes*. As we shall see, this anthropological Archimedian point grounds both aspects of Kierkegaard's method, the cognitive approach as well as the experimental/psychological approach to philosophy, as a human or spiritual science.

Let us now focus attention upon these two aspects of the overall method. We shall begin with an analysis of the elements inherent in the cognitive approach.

"The cognitive approach to method," although it is grounded in the extra-methodological assumption of *unum noris omnes*, takes its *operational* launching point from yet another Kierkegaardian motif, the principle of "consistent thinking."/29/ Consistent thinking meant for Kierkegaard a rigorous and thoroughly scholarly commitment to a "comprehensive view of life's manifold forms, especially of human existence, with an emphasis on subjective actuality and its relationship to Christianity as the base."/30/ Consistent thinking, then, lies at the root of all logic and mathematics. As the root of apodicticity, it became for Kierkegaard the model for all scholarly treatment of other subjects as well. Kierkegaard's most visible use of consistent thinking is his extensive use of indirect proof, a technique exhaustively employed in *Begrebnet Angest*.

Gregor Malantschuk sees consistent thinking's operational function as follows:

> The ideal in all spheres of scholarship is to reach a scientifically and scholarly correct treatment of the subjects investigated. An ideal achieved only by way of thorough consistency. The two auxiliary disciplines which must be considered in this connection are logic and mathematics, the whole structure of which rests on the principle of consistency. Consistency, or more accurately, consistent thinking, is the implicit but indispensible premise for these two disciplines; consequently they become the norms for a consistent and thereby scholarly treatment of other areas of scholarship./31/

To merit the designation of "scholarly activity," indirect proof (as well as direct proof), must follow the demands of consistent thinking. Otherwise what would follow is "fantastic thought" (i.e., that which is grounded in fantasy or which fantastically explains everything)./32/

Kierkegaard's basis for this cognitive approach came from his reading of F. C. Sibbern's work, *Om Erkjendelse og Granskning* [*On Knowledge and Research*]./33/ Sibbern helped Kierkegaard to define the parameters of his cognitive approach and to apply this approach in three stages: First, with regard to his general human sciences project, Kierkegaard could work his way through existing solutions to problems within the humanistic disciplines of his project. Only then could he arrive at verification of these solutions. Kierkegaard could construct fresh and correct viewpoints on these problems by consistently thinking through their implications. Finally, Kierkegaard could apply the verified conclusions to the disciplines within his fundamental project./34/ All this was possible only through the power of consistent thinking as the first element of his cognitive approach to method.

Another element of Kierkegaard's cognitive approach was also inherited from Sibbern. This was the notion of *collateral* or *parallel development* of the "individual's" mental-spiritual being. During 1838, Kierkegaard read and absorbed Sibbern's work, *Bamaerkninger og Undersøgelser, fornemmelig betraeffende Hegels Philosophi, betraget i Forthold til vor Tid* [*Observations and Investigations Particularly on Hegel's Philosophy, Seen in Relation to Our Time*]./35/ From this work, he gained the following essential insight: ". . . the development of the mental-spiritual content in the realms of both external and internal actuality takes place *simultaneously* along many lines."/36/ (Italics added.)

While Sibbern thought there were many different specific collateral concepts, indeed as many as there are types of existential "categories" or mental-spiritual qualifications,/37/ there are, he claimed, "The genuine synthetic unities in a special sense or the *genuine* synthetic combinations of two forming a third . . ."/38/ and "such triads as we have, for example in *cognition, feeling*, and *will*."/39/ Kierkegaard did not agree, however, that *all* synthetic unities were co-equal. Rather, as we shall see in our discussion of *Begrebnet Angest*, he believed that one of each pair dominated the other. Thus in *Begrebnet Angest*, for example, the eternal dominates the temporal, freedom dominates necessity, and so on. He did, however, concur with Sibbern that cognition, feeling, and will were indeed co-equal./40/

Kierkegaard next worked out the specific existential "categories" most frequently employed in *Begrebnet Angest*: namely the temporal-eternal, necessity-freedom, finite-infinite, and body-mind-spirit. Moreover, Kierkegaard uses the concepts of being-essence and quality-quantity frequently throughout the corpus of his writings. These he overlaid on what he saw as the fundamental categories of human being: understanding, feeling, and will, resulting in a grid which displays the basic existential structure of the individual./41/ This process is summarized by Kierkegaard in the *Journals*:

> If the understanding [*Forstanden*], feeling, and will are essential qualifications in a man, belong essentially to human nature, then all this chaff that the world development now occupies a higher level vanishes into thin air, for if there is a movement in world history, then it belongs essentially to providence, and man's knowledge of it is highly imperfect. . . .
>
> The great individual is great simply because he has everything at one. . . . [He] is not thereby different from the insignificant individual by possessing something essentially different or by having it in another form . . . but by having everything to a greater degree./42/

Armed then with consistent thinking and the principle of collateral development as cognitive weapons, Kierkegaard was prepared to attempt an understanding and interpretation of human existence in the light of Christianity.

Kierkegaard believed the existential categories (or qualifications of the human spirit) could be explicated and described only from a decidedly Christian perspective. He specifically thought that the categories or qualifications of innocence-guilt, finite-infinite, faith-doubt made sense only in relation to Christianity in general and man's Christian spirit in particular. Therefore, the world and human existence can be grasped only from the Christian world view. According to Kierkegaard, "the true Christian view, [is] that universally human existence does not explain Christianity and that Christianity is not simply another factor in the world, but that Christianity explains the world."/43/

The phenomenon of Angst ultimately makes possible the individual's appropriation of the Christian understanding of the world. But to attain such understanding, the individual must begin his collateral development from a non-Christian, totally aesthetic point of departure. Then, through a series of qualitative leaps or individual commitments, he can rise to the Christian understanding of the world. Gregor Malantschuk sees this conceptual framework within the purview of *Begrebnet Angest*. His comments are worth repeating here because they capture the overall task of Kierkegaard's intent in writing the work on

Angst. Malantschuk observes:

> The single individual must constantly begin his development from below and from the existential point of view, must go through a prolonged process before he has the courage and earnestness necessary to make the decision regarding Christianity. . . . Virgilius Haufniensis in *The Concept of Anxiety* [*Begrebnet Angest*] takes upon himself the task of depicting the mental-emotional aspect of man . . . [and] points unflaggingly to the fact that the actual transition from these mental-emotional positions to a new position which Christianity offers can occur only by a qualitative leap./44/

We may conclude our discussion of Kierkegaard's cognitive approach to method by observing that Kierkegaard was anything but an undisciplined thinker. Rather, the essence of his cognitive approach is a rigorous application of logical canons to consistently thinking through the implications of any given perspective on the "existential categories" or qualifications of the human spirit. Nevertheless, to interpret Kierkegaard's thought only from this perspective is to miss his thrust completely. Running collaterally to the cognitive is a dialectically opposed approach, one grounded in the human drama: "the experimental/psychological approach" to method.

The experimental/psychological approach. We have suggested that the experimental/psychological approach to method is Kierkegaard's dialectic and experimental counterpart to the cognitive approach. *But both approaches are united at root in the general anthropological assumption of* unum noris omnes *(if you know one, you know them all).* How is this so? Kresten Nordentoff provides the clearest answer to this question.

> In the final analysis, all understanding is self understanding. An "objective" understanding of one's surroundings, which is not grounded in self-understanding, is either triviality or a demonic illusion. Only by means of self-understanding can a genuine understanding of one's surroundings come into being. Therefore, self-understanding is the *conditio sine qua non* for the areas of knowledge which are under discussion here—and therefore, Kierkegaard's psychology is based upon self-analyses./45/

Thus Kierkegaard's psychology, as well as the logic of consistent thought, begins in self-introspection: the apodictic center of our knowledge of others. Moving from self-knowledge to knowledge of others requires that we verify our conclusions about the other on the basis of the *unum noris omnes* principle: self-knowledge can always be *deceptive*, of course, regardless of how cognitively rigorous one's thought is regarding the "existential categories" or qualifications of the human spirit.

This problem is solved by Kierkegaard in his experimental method of observing others to confirm or refute conclusions drawn from self-knowledge. With this in mind, a quote cited above takes on new meaning.

> . . . every person possesses in himself, when he looks carefully, a more complete expression for everything human than the *summa summarium* of all knowledge he acquires by that method [i.e., by learned studies]./46/

This means:

> . . . what holds here in a profound psychological decision is *unum noris omnes*. When the possibility of sin is shown in one, it is shown in all, and all that is left to the ideal arena of observation are considerations of more or less./47/

The experimental/psychological approach, then, is founded in Kierkegaard's self-understanding. From such understanding Kierkegaard was able to establish his unique perspectives on the "existential categories" of being human. Such perspectives as hypotheses about the meaning of being human require confirmation, according to Kierkegaard, in "real life" experience—that is, in the marketplace, at church meetings, or at the theater. Only through *empathetic insight* and the uncovering of the hidden and deeper dimensions of the subject's spirit could such observations be carried out. To accomplish these ends, therefore, the psychologist must get beyond the illusory veil of merely observing his subject. Rather, the hidden and deeper well-springs of the spirit are revealed only when the psychologist *closely and empathetically listens* to the subject. Psychological observation is therefore an auditory rather than visual mode of observation./48/

But according to Kierkegaard, a purely clinical, objective, and disinterested approach to method *is not enough* in genuine psychological observation. Indeed, precisely because clinical knowledge *is* disinterested, nothing is revealed of the passionate and frequently demonic feelings within the subjects's spirit. The psychologist must listen in silence for the spirit's dark and hidden passion to slip "out to chat with itself in the artificially constructed nonobservance and silence." This quest for knowledge of the subject's spirit, therefore, requires *interested* rather than disinterested knowledge.

So the more the psychologist is interested in the subject, the more sound and profound will be his knowledge of the subject. Thus in 1844 Kierkegaard wrote in his *Journal*: "Psychology is what we need, and above all, expert knowledge of human life and sympathy with its interests. Thus, here there is a problem the solution of which must precede any talk of rounding out a Christian view of life."/49/ This passage shows the intimate and even necessary connection between psychology and Kierkegaard's ultimate goal of articulating man's existential qualifications, particularly with regard to the Christian view of life.

As we indicated above, this psychology, rather than being objective and disinterested, is *passionately interested* in the human subject. On this point Kierkegaard parted company with his mentor in the cognitive approach to method, F. C. Sibbern. Kierkegaard remarked that Sibbern was satisfied with simply *describing* human emotions in a dispassionate way. As a result, says Kierkegaard, Sibbern's approach to psychology lacks "an eye for the disguised passions, for the reduplication by which one passion takes the form of another."/50/ Bearing this passionate, fundamentally interested quest for confirmation in mind, Nordentoft summarizes economically the role of the master psychological observer as well as the validity of his conclusions:

> That the requirement of responsible self-understanding is both existentially and scientifically valid can . . . be seen in Climacus' discussion of "the subjective thinker" in *Concluding Unscientific Postscripts*. First (and last), one learns that this thinker must have fantasy and feeling; he must be able to talk poetically and ethically, and he must be able to think dialectically. But, especially, *he must have passion*./51/ (Italics added.)

If the preceding is so, then Angst could not have sprung full-blown into Kierkegaard's philosophical lexicon. Indeed, as we suggested at the end of the section on influences, it may be more fruitful to understand and interpret Angst in Kierkegaard's thought if we start from the following hypothesis: the phenomenon of Angst became a passionate interest for Kierkegaard based upon his feelings regarding his father, Michael, and his rejected fiancee, Regine Olson. Too, on a more cognitive plane, Angst may have arisen as a result of Kierkegaard's early family encounter with Christianity and its rigorous ethical demands. As the literature shows, the young Kierkegaard could deal with such demands intellectually, but not emotionally. This interpretation is at least partially grounded in the following *Journal* entry:

> If I had had faith, I would have stayed with Regine. Thanks to God I now see that . . . I would have married her—there are so many marriages which conceal little stories. That I did not want, then she would have become my concubine; I would rather have murdered her.—But if I were to explain myself, I would have had to initiate her into terrible things, my relationship with my father, his melancholy, the eternal night brooding within me, my going astray, my lusts and debauchery, which, however, in the eyes of God are perhaps not so glaring; for it was, after all, anxiety [Angst] that made me go astray, and where was I to seek a safe stronghold when I knew or suspected that the only man I had admired for his strength was tottering./52/

Through the *unum noris omnes* principle, Kierkegaard later concluded that Angst is a basic pre-disposition—a necessary condition of the human spirit. Angst is, therefore, a basic "existential category" or a qualification of the human being's subjective actuality: A conclusion grounded in the apodicticity of self-analysis, and one experimentally confirmed by continuous experimental/psychological observations of his father's enormous melancholic Angst. Specifically, Kierkegaard employed those means that were to become the experimental/psychological approach to method so as to wrest his father's dreadful secret from him. The factual existence of Michael Kierkegaard's dreadful secret and the Angst associated with it, when seen from the standpoint of the *unum noris omnes* principle, revealed the primordial pre-disposition of Angst to be a universal human phenomenon.

Thus Angst required both consistent thought and further experimental/psychological observation to confirm earlier conclusions regarding its staggering power over not only his father but, through the *unum noris omnes* principle, human subjectivity in general. Only later, as Kierkegaard worked through the writings of Boehme, von Baader, Schelling, and Hamann, was he able to pull together in his unique existential synthesis his father's melancholic Angst, his own Angst (which made him go astray) and the primordial Angst of which Boehme and Schelling spoke—all to see Angst's essential role in human existence as a pre-reflective apprehension of the infinite freedom of possibility.

To ground this interpretation as well as to develop it collaterally and organically, we must more closely examine the way Kierkegaard consistently thought through his early experience of Angst. It is clear that his mature interpretation of Angst is the result of several *interlarded stages* of understanding and interpreting Angst's power.

Presentiment as the Immediate Horizon for Angst

Kierkegaard's relationship to his father, Michael P. Kierkegaard, has been the subject of much discussion in connection with the phenomena of melancholy and Angst. The basis for such discussion appears to stem, at least in part, from a previously cited *Journal* entry, composed after Kierkegaard has broken off his engagement to Regine Olson. Let us return to this passage in the light of this new disclosure. Again, we quote:

> I would have married her—there are so many marriages which conceal little stories . . . But if I were to explain myself, I would have had to initiate her into terrible things, my relationship to my father, his melancholy, the eternal night of brooding within me, my going astray, my lusts and debauchery which, however, in the eyes of God are perhaps not so glaring; for it was, after all, anxiety (Angst) that made me go astray, and where was I to seek a safe stronghold when I knew or suspected that the only man I had admired for his strength was tottering.

The haunting echo of the Angst that made the young Søren "go astray" when he could find no safe stronghold in his father shows the obvious personal Angst of Søren Kierkegaard in the face of his father's melancholy. It would neither be accurate nor at all fruitful to argue that Kierkegaard derived the concept of Angst directly from his father's personal melancholy. However, it could well be argued that such melancholy may have provided the existential conditions that led Kierkegaard to develop consistently an intermediate set of concepts *between* melancholy and Angst. Specifically, at an early age Kierkegaard set out to appropriate and consistently think through the romantic idea of the "master-thief" as a means of "stealing" his father's secret from him, thus forcing into the open what was concealed in Michael Kierkegaard's melancholic inclosing reserve [*det Indesluttede*]./53/

So Kierkegaard's desire to pry open his father's secret, to expose the reasons for his father's melancholy, gave birth to several romantically wistful *Journal* entries concerning the "master-thief." The master-thief, says Kierkegaard, ". . . is not a man who tries to lead others astray; on the contrary, he dissuades them from leading such a life."/54/ This desire to heal his father's melancholy led Kierkegaard from the master-thief idea to the discovery of a broader psychological phenomenon, one which was capable of uncovering the hidden. *This was the phenomenon of "presentiment."*

In particular, Kierkegaard came to believe that certain presentiments revealed hidden psychic motivations which held the individual in the demonic power of inclosing reserve. The discovery of presentiment appears to have surfaced in 1837, eight years before *Begrebnet Angest* was written:

> A certain presentiment seems to precede everything which is to happen . . . ; but, just as it can have a deterring effect, it can also tempt a person to think that he is, as it were predestined; he sees himself carried on to something as though by consequences beyond his control. Therefore one ought to be very careful with children, never believe the worst and by untimely suspicion or by a chance remark . . . occasion an anguished consciousness in which innocent but fragile souls can easily be tempted to believe themselves guilty, to despair, and thereby to make the first step toward the goal foreshadowed by the unsettling presentiment—a remark which gives the kingdom of evil, with its stupefying, snakelike eye, an occasion for reducing them to a kind of spiritual paralysis. Of this, too, it may be said: woe unto him by whom the offence comes./55/

In the margin of this entry, Kierkegaard scribbled: "The significance of topology with reference to a theory of presentiment."/56/ In this passage is revealed the immediate predecessor of Angst as Kierkegaard understood that phenomenon in 1844. We may ground this interpretation by reference of Kierkegaard's own words in *Begrebnet Angest*: ". . . the nothing of anxiety [Angst] *is a complex of presentiments*, which, reflecting themselves in themselves, come nearer and nearer to the individual, even though again, when viewed essentially in anxiety, they signify a nothing . . ." (italics added)./57/

Thus, while presentiment is not identical with Angst, it is linked intimately with Angst. This link, we suggest, provided Kierkegaard the key to a more mature understanding of the Angst phenomenon itself. How is this so? To answer this question, we must observe more closely the phenomenon of presentiment in Kierkegaard's thought.

As was seen above, presentiment means an apprehension preceding "everything which is about to happen." But presentiment cannot be anything like a conceptually transparent prevision. On the contrary, presentiment is so dark and murky that it fills the soul with Angst. Says Kierkegaard:

> All presentiment is murky and rises all at once in the consciousness or so gradually fills the soul with anxiety [Angst] that it does not arise as a conclusion from given premises but always manifests itself in an undefined something; however, I now believe more than ever that an attempt should be made to point out the subjective pre-disposition and not as something unsound and sickly, *but as an aspect of normal constitution.*/58/ (Italics added.)

Here we see presentiment is a normal and subjective *predisposition* wherein one experiences Angst in the face of "an undefined something." To be sure, this *early* discussion of presentiment lacks the usual rigor and clarity of Kierkegaard's consistent thinking. But as with the *phenomenon* of presentiment, the *notion* of presentiment was emerging gradually to Kierkegaard as a theme for philosophical investigation.

Not until September 10, 1839, could Kierkegaard more clearly express his view of presentiment. On that date in his *Journal*, not only did Kierkegaard define presentiment, but he describes its horizons with post-Husserlian phenomenological rigor.

> Presentiment is not linked to the direction of the eye's orientation toward existence and its future but to the reflex of the eye's direction toward the past so that the eye, by staring at what lies behind it (in another sense, ahead of it) develops a disposition to see what lies ahead of it (in another sense behind it).

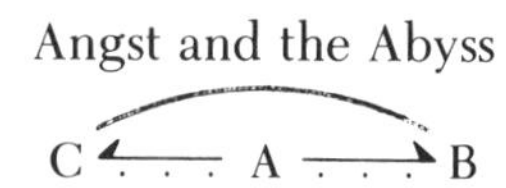

> For example, if A is the present, the time in which we are living, and B the future, then it is not by standing in A and turning my face toward B that I see B; for by turning thus I see nothing at all, but when C is the past, then it is by turning toward C that I see B, just as, in fact, in Achim V. Arnim's novel the presentiment eyes of Alrunen were situated *in the back of his head*, whereas his other two eyes, which were farsighted in the ordinary way—that is, two regular eyes—were in his forehead just like the eyes of other men, or that part of the head turned toward the future./59/

Thus by 1839 Kierkegaard was able to articulate presentiment as a temporally orienting phenomenon in which the future is apprehended by reference to the past. As such, presentiment has a close relationship to the category of possibility; that category under which the phenomenon of Angst would later be discussed in *Begrebnet Angest*.

In summation, then, in Kierkegaard's thought, presentiment has several different facets:/60/ (1) "historical presentiments," which arise when the individual absorbs himself in history's driving forces; (2) "prophetic presentiments," which arise when these historical forces are interpreted from within a religious of "prophetic" perspective; (3) "personal presentiments," which arise when the individual contemplates his own destiny; and finally (4) "psychological presentiments," which arise when the experienced observes "on the basis of a person's single utterance . . . a survey of all the consequences of his lifetime."/61/

While the first two of these four may be said to closely resemble the cognitive element of Kierkegaard's thought, the other two are related to his experimental/psychological thought. Moreover, and equally important, the latter may even have been derived from the former, since in Kierkegaard's own personal life his earliest personal and psychological presentiments were the result of his relationship to his father's dreadful secret. This became, in our view, the basis for the master-thief idea which in turn made possible the theory of presentiment. As was shown above, the first clear recognition of the phenomenon of presentiment was entered into the *Journal* in 1837. But during the same year, Kierkegaard entered another entry describing presentiment, this time in relation to the cognitive and dogmatic problem of original sin. We are now nearing *Begrebnet Angest*. Kierkegaard notes:

> For something to become really depressing there must be first of all, in the midst of all possible favors, a *presentiment* that it might just be all wrong; one does not become conscious of anything very wrong in himself, but it may lie in the familial context; then *original sin* displays its consuming power, which can grow into despair and have a far more frightful effect than the particular whereby *the truth of the presentiment is verified.*/62/

The final transition from the theory of presentiment and original sin to Angst and original sin had to wait until 1842, two years prior to *Begrebnet Angest*'s publication. Kierkegaard says:

> The nature of original sin has often been explained and still a primary category has been lacking—it is anxiety [Angst]; this is the essential determinant. Anxiety is a desire for what one fears, a sympathetic antipathy; anxiety is an alien power which grips the individual, and yet one cannot tear himself free from it and does not want to, for one fears, but what he fears he desires. Anxiety makes the individual powerless, and the first sin always occurs in weakness; therefore it apparently lacks accountability, but this lack is the real trap./63/

At last we have arrived at *Begrebnet Angest*'s doorstep.

The Concept of Angst in Begrebnet Angest

Preliminary Considerations

We move now from the micro-hermeneutical horizons to Kierkegaard's specific Angst depiction given in *Begrebnet Angest*. The path for this transition has been paved, it is hoped, by a somewhat detailed account of Kierkegaard's cognitive influences as well as his personal encounter with Angst in the face of his relationships to his father and Regine Olson. These relationships allowed Kierkegaard to discover "ontic Angst," and much more importantly, that discovery led Kierkegaard to see a deeper ontological meaning of Angst: something like a primordial predisposition of the human spirit./64/ Certainly Kierkegaard did not use the expression "primordial pre-disposition," but the following quotation shows he understood well enough the ontological significance of Angst. As early as 1837, Kierkegaard wrote:

> Deep within every human being there still lives the anxiety [Angst] over the possibility of being alone in the world, forgotten by God, overlooked among the millions and millions in this enormous household. A person keeps this anxiety at a distance by looking at the many found about who are related to him as kin and friends, but the anxiety is still there, nevertheless, and he hardly dares think of how he could feel if all this were taken away./65/

Two years later Angst's ontological significance touched Kierkegaard at even a deeper level: "All existence [*Tilvaerelsen*]/66/ makes me anxious, from the smallest fly to the mysteries of the incarnation; the whole thing is inexplicable to me, I myself most of all. My distress is enormous, boundless; no one knows it except God in heaven, and he will not take compassion on me."/67/

Thus we see that by 1839 Kierkegaard had already attained remarkable breadth and depth in his reflections on the ontological significance of Angst. It remained for him to couch these insights into the framework of his overall project: the rigorous examination of the "existential categories" or qualifications of the human spirit, consistently thought through, tested, and verified through the experimental/psychological approach. *Begrebnet Angest* is a part of that framework: an outstanding example of the *unum noris omnes* principle employed as a rigorous analysis of Angst from both the cognitive and the experimental/ psychological approaches.

Before turning to *Begrebnet Angest* itself, we feel compelled to state that it is not our purpose here to provide a page-by-page commentary on *Begrebnet Angest*. Further, it is not our task to explore each of the subtle psychological insights into Angst offered in that work., In our view, Kresten Nordentoft/68/ has addressed this problem with insightful precision, and we see no reason to duplicate his effort.

Thus we turn now to Angst in *Begrebnet Angest*, bearing in mind that we are dealing not only with Søren Kierkegaard, but also with his pseudonymous psychologist, Virgilius Haufniensis, "the watchman of Copenhagen." This is despite the fact that Kierkegaard considered publishing it under his own name./69/

Begrebnet Angest—*The Cognitive Dimension*

We would do well to begin our discussion by asking a basic question: namely, why did Kierkegaard-Haufniensis/70/ write

Begrebnet Angest in the first place? The answer should provide at least one set of guidelines for interpreting Angst in Kierkegaard-Haufniensis' work. Stating our conclusion first, we believe Kierkegaard-Haufniensis wrote the work largely as polemic against Hegel. This is shown in the opening lines of the book's "Introduction" where the author states his purpose: to show that Hegel's "system" not only fails to recognize the integrity of the human sciences [*Geisteswissenschaften*] but it robs thought of its proper rigor, leading to fantastic and murky results. Specifically Kierkegaard-Haufniensis states on the opening page of *Begrebnet Angest*:

> The view that every scientific issue within the larger compass of science has its definite place, its measure and its limit, and thereby precisely its harmonious blending in the whole as well as its legitimate participation in what is expressed by the whole, is not merely a *prim desiderium* [pious wish] . . . not merely a sacred duty . . . , it also serves the interest of every more specialized deliberation, for when the deliberation forgets where it properly belongs, . . . it forgets itself and becomes something else, and thereby acquires the dubious perfectibility of being able to become anything and everything. (*CA*, p. 9)/71/

For Kierkegaard-Haufniensis, Hegel's system had indeed become "anything and everything." Thus, one principal task of *Begrebnet Angest* was to correct Hegel's error with respect to two disciplines within the human sciences: dogmatics and psychology./72/ After showing how an uncritical acceptance of the *Aufheben* (annulment) principle leads to conceptual confusion in both logic and dogmatics, Kierkegaard-Haufniensis spells out the specific thesis of *Begrebnet Angest*:

> The present work has set as its task the psychological treatment of "Anxiety [Angst]," but in such a way that it consistently keeps *in mente* [in mind] and before its eye the dogma of Hereditary Sin. Accordingly, it must also, although tacitly so, deal with the concept of sin. Sin however, is no subject for psychological concern. . . . (*CA*, p. 14)

From the beginning, therefore, Kierkegaard-Haufniensis distinguishes between two conceptual frameworks within *Begrebnet Angest*: (1) the phenomenon of Angst as a concern of psychology,/73/ and (2) the problem of "hereditary sin" as a concern of dogmatics. From the standpoint of consistent thinking, therefore, psychology must be confined to illuminating Angst as the pre-condition of hereditary sin. Thus, Angst can never *explain* hereditary sin itself.

Kierkegaard's point of departure, that psychology must confine itself to showing the pre-conditions of hereditary sin, circumscribes very neatly the role of Angst in *Begrebnet Angest*. We will attempt therefore to remain within this circumscription so as to avoid the difficult problem of hereditary sin, which, strictly speaking, is beyond our scope of concern here. From the cognitive perspective, then, Angst first emerges as the pre-condition of hereditary or original sin. We must now inquire into how Kierkegaard-Haufniensis describes this pre-condition theologically.

Prior to Adam's sin, the first sin as well as the Fall of man, primal man was in a dreaming state./74/ As a dreaming spirit primal man was totally ignorant of sin:

> In this state there is peace and repose, but there is simultaneously something else that is not contention and strife, for there is indeed nothing against which to strive. What, then, is it? Nothing. But what effect does it have? It begets anxiety [Angst]. This is the profound secret of innocence, that it is at the same time anxiety. Dreamily the spirit projects its own actuality, but this actuality is nothing, and innocence always sees this nothing outside itself. (*CA*, p. 41)

Hence, Angst is first and foremost a pre-reflective apprehension of the Nothingness outside the innocent spirit's dreamlike state. Nevertheless, says Kierkegaard-Haufniensis, we cannot deny Angst's existence. Angst is pre-reflective because the innocent spirit has no "self-consciousness" of sin yet. It is apprehensive because as a basic pre-disposition Angst is experienced directly. Yet the spirit senses no direct threat as, for example, would be the case with fear. Precisely upon these grounds Kierkegaard-Haufniensis provides his classical distinction between Angst and fear, a distinction that is almost universally agreed upon today.

> The concept of anxiety [Angst] is almost never treated in psychology. Therefore, I must point out that it is altogether different from fear and similar concepts that refer to something definite, whereas anxiety is freedom's actuality as the possibility of possibility./75/ (*CA*, p. 42)

In this quotation is the first explicit definition of Angst: Angst is psychologically freedom's actuality as the possibility of possibility. Much of what follows in *Begrebnet Angest* seeks to clarify the significance of this pregnant concept.

This psychological definition of Angst is supplemented dialectically by a more philosophical (Hegelian) definition that became the *locus classicus* of Kierkegaard-Haufniensis' Angst concept:

> When we consider the dialectical determinations of anxiety [Angst], it appears that exactly these have psychological ambiguity. Anxiety is *sympathetic antipathy* and *an antipathetic sympathy*. One easily sees, I think, that this is a psychological determination in a sense entirely different from the *concupiscientia* [inordinate desire] of which we spoke. Linguistic usage confirms this perfectly. One speaks of a pleasing anxiety, a pleasing anxiousness [*Beaengstelse*], and of a strange anxiety, a bashful anxiety, etc. (*CA*, p. 42)

This characterization may be compared with the 1842 *Journal* entry in which the italicized English words above are made somewhat clearer.

> The nature of original sin [hereditary sin] has often been explained and still a primary category is lacking. This is anxiety [Angst]; this is an essential determinant. Anxiety is *the desire for what one fears*, a sympathetic antipathy, anxiety is an alien power which grips the individual, and yet one cannot tear himself free from it and does not want to, for one fears, but what he fears he desires. Anxiety makes the individual powerless, and the first sin always occurs in weakness; therefore, it apparently lacks accountability, but this lack is the real thing./76/ (Italics added)

Thus from the cognitive perspective of Kierkegaard's thought, we may say that sympathetic antipathy and antipathetic sympathy are the dialectical poles of Angst. Or as Walter Kaufmann has said, Angst ". . . involves a deep ambivalence. We are divided against ourselves and want something that another part of us does not want."/77/

Angst is also couched in another dialectic, that between objective Angst and subjective Angst. Kierkegaard-Haufniensis does not give much space to objective Angst, and if Vincent McCarthy is right, this is because in order to do so, subjective

Angst must have a conceptual counterpart from which it can be distinguished—objective Angst./78/ It is certainly clear that objective Angst claims only four pages of *Begrebnet Angest*, and much of that is taken up in a lengthy footnote criticizing Schelling in this use of Angst as a concept./79/ Subjective Angst, however, claim almost two-thirds of *Begrebnet Angest*, a sizable difference to say the least.

Nevertheless, let us examine these two Angst dialectical poles. Objective Angst for Kierkegaard is a quantitative rather than qualitative phenomenon; it is the total amount of Angst in creation since Adam's first sin. Objective Angst therefore is grounded in Adam's original sin and is expressed as ". . . the reflection of the sinfulness of the generation in the whole world" (*CA*, p. 57). The effect of sin in non-human existence [*Tilvaerelse*] Kierkegaard-Haufniensis calls objective Angst (*CA*, p. 57), concretely displayed in Romans 8:19, "the eager longing of creation" (*CA*, p. 58). What Kierkegaard-Haufniensis means by reference to this biblical phrase is as follows: expressions such as "longing," "expectation," etc., imply a preceding state within the individual from which he longs to be delivered. This feeling, this longing, is primordial Angst, proclaiming itself mutely but powerfully. Simple longing alone is not sufficient to deliver the individual from the preceding state.

Subjective Angst, on the other hand, is described as an internal condition within primal man. Kierkegaard-Haufniensis says:

> Anxiety [Angst] may be compared with dizziness. He whose eye happens to look down into the yawning abyss becomes dizzy. But what is the reason for this? It is just as much in his own eye as in the abyss, for suppose he had not looked down. Hence anxiety is the dizziness of freedom, which emerges when the spirit wants to posit the synthesis and freedom looks down into its own possibility, laying hold to finiteness to support itself. Freedom succumbs in this dizziness. (*CA*, p. 61)

So with subjective Angst we return again to the metaphor of the abyss. But the abyss alone does not make the spirit dizzy. Rather it is the spirit which makes dizziness possible to begin with as man stares at the possibilities before him. This is true not only for Adam as primal man but for every subsequent individual as well.

Thus at the edge of the abyss stands the dizzy individual

grasping at the finite ontic world for support in the face of the abyss of ontological possibility. Then comes the leap. "In that very moment everything is changed," says Kierkegaard-Haufniensis, "and freedom, when it again rises [from the abyss] sees that it is guilty. Between these two moments lies the leap, which no science has explained and no science can explain" (*CA*, p. 61). We may summarize this dialectical process in the following way: At the outset, Adam as primal man, was in a dreamlike state of innocence. But contained within his composite being, a being composed of the psyche and the body synthesized in spirit (*CA*, p. 85), is the pre-reflective apprehension of freedom as an ontological condition of being human. The more such a pre-reflective state becomes *reflective*, the more possibility raises itself as the specter of non-actuality. Possibility becomes literally non-actuality or a "no-thingness." At the moment of this conscious reflection, the spirit stands at the face of the abyss peering down into the bottomless chasm of its own possibilities. The individual's infinite possibilities are both attractive and frightening: Attractive because any one of them can potentially be actualized; and frightening because when one *is* actualized, all others must be abandoned. Thus a specific existence mode comes into being for the individual; *he has made an existential choice*. Hence all other possibilities must be precluded at least for the moment. But just because the choice is made, this does not quell Angst. On the contrary, Angst is always there, as long as the individual has temporally before him an infinite number of possibilities; Angst is there, in other words, as long as he exists.

For Adam as primal man, this means that he innocently and dizzily stared down into the abyss of ontological possibility. In his dizzy state, he *leapt* into that abyss, thereby actualizing the possibility of sin through a concrete act of choice. Adam emerged from the abyss in a new existence medium, namely, ontological guilt. Thus in this extended metaphor, the fall of man is interpreted as the leap into the abyss of ontological possibility as an intentional act of freedom. Note that none of this *explains* hereditary sin. It does not, indeed *cannot*, deal with the question of *why* Adam took the leap. That question, says Kierkegaard-Haufniensis, is clearly germane only to the realm of dogmatics and not the psychological. The latter realm, psychology, can only inquire into the subjective pre-condition of sin, namely, the Angst of dizzying freedom.

Still within the framework of the cognitive aspect of his method, Kierkegaard-Haufniensis explores the "existential categories" or the qualifications of the human spirit from the standpoint of Angst. Just as man's existential categories include the synthesis of the psyche and the body, says Kierkegaard-Haufniensis, they also include those of the temporal and the eternal (*CA*, p. 85). Just as spirit is the third factor in this first synthesis, "the moment"/80/ is the third factor in the temporal-eternal polarity. By the concept of "the moment," Kierkegaard-Haufniensis means the positing of the eternal in *the present*. This does not imply the positing of the eternal in *time*, for that is the ultimate paradox of Christianity (i.e., Eternal God becoming the temporal Jesus Christ) much as "the present" is a category of time. Thus "the moment" is an atom of the eternal in man which corresponds to man's eternal spirit. Says Kierkegaard-Haufniensis: "The synthesis of the temporal and the eternal is not another synthesis but is the expression of the first synthesis according to which man is a synthesis of psyche and body that is sustained by spirit. As soon as the spirit is posited, the moment is present" (*CA*, p. 88).

But what does all this have to do with Angst? It will be recalled that subjective Angst is related to possibility. Put in this present context, "the moment" corresponds to the future. Thus the spirit projects itself into the future-moment when it seeks to *actualize* one of its infinite possibilities. Hence Angst ontologically is the feeling arising in the spirit as it leaps into the futural moment. Kierkegaard-Haufniensis observes:

> Just as . . . the spirit . . . when it is about to posit the synthesis as the spirit's (freedom's) possibility in the individuality, expresses itself as anxiety [Angst], so here the future in turn is the eternal's (freedom's) possibility in the individuality expressed as anxiety. As freedom's possibility manifests itself for freedom, freedom succumbs and temporality emerges. . . . (*CA*, p. 91)

From these descriptions and analyses Kierkegaard-Haufniensis observes: "the possible corresponds exactly to the future. For freedom, the possible is the future, and the future is for time the possible" (*CA*, p. 91). So, the author concludes: "an accurate and correct linguistic usage therefore associates anxiety [Angst] with the future" (*CA*, p. 91).

In the previous sections we saw that the temporal-eternal is but one of the "existential categories" or qualifications of the human spirit which must be developed collaterally with others such as necessity-freedom, and the finite-infinite. In the course of consistently thinking through Angst in *Begrebnet Angest* each of these are developed,/81/ thus rounding out the task of Kierkegaard's cognitive approach to philosophical investigation. We may conclude this subsection by observing that with one half of our discussion of *Begrebnet Angest* now behind us, there remains the task of interpreting Angst from the standpoint of Kierkegaard's experimental/psychological approach.

Begrebnet Angest—*The Experimental/Psychological Dimension*

As we saw above, Kierkegaard's experimental/psychological approach is the dialectical counterpart to his cognitive approach. We might reasonably expect, therefore, that *Begrebnet Angest* would discuss Angst from the experimental/psychological approach if these two modes are of equal importance to his thought. As we shall see, Kierkegaard-Haufniensis uses the experimental/psychological approach with such rigor and clear thinking that many of his psychological observations, although they pre-date Freud and the psychoanalytic movement, are still relevant in understanding Angst and its important relationship to neurosis. The essential chapter in which this dimension is revealed is Chapter IV of *Begrebnet Angest*, entitled "Anxiety [Angst] of Sin or Anxiety as a Consequence of Sin in the Single Individual."

From one perspective, the discussion offered in that chapter is a synthesis of the cognitive and the experimental/psychological approaches; for the "existential categories" of "good" vs. "evil" in relation to Angst are thoroughly thought through. But in the broader context we believe that Kierkegaard's Chapter IV is an experimental/psychological contribution first and foremost, and a cognitive one secondarily. Why is this so? We answer, because its *basic thrust* is a discussion of inclosing reserve [*det Indesluttede*]/82/ as the *essence of the demonic*—the very problem which the young Kierkegaard suspected was at the core of his father's melancholy, and as the man was the key to his ontic Angst in relation to Regine Olson.

Chapter IV begins, then, with a discussion concerning Angst

and Evil vs. Angst and Good. While Angst about Evil is revealed in repentance, a kind of repentance, moreover, that stands before the qualitative leap, Angst about the Good/83/ is revealed in the inclosing reserve of the demonic. The demonic is a term describing a condition wherein the individual *wants* to remain unfree. Specifically, the individual wants to remain in the bondage of sin on a permanent basis. Hence, his Angst becomes a profound terror of becoming good,/84/ of becoming free. The phenomenon of the demonic, says Kierkegaard-Haufniensis, first appears in the New Testament in Matthew 8:28–34; Mark 5:1–20; Luke 8:26–39; as well as Luke 11:14. According to Matthew, to cite one example, the demonic is demonstrated as follows:

> When he [Jesus] reached the other side [of the sea of Galilee] in the country of the Gadarenes, he was met by two men who came out from the tombs; they were possessed by devils, and so violent that no one dared pass that way. "You son of God," they shouted, "what do you want with us? Have you come here to torment us before our time?" In the distance a large herd of pigs was feeding; and the devils begged him: "If you drive us out, send us into that herd of pigs." "Begone!" he said. Then they came out and went into the pigs; the whole herd rushed over the edge into the lake, and perished in the water.
>
> The men in charge of them took to their heels, and they told the whole story and what had happened to the madman. Thereupon all the town came out to meet Jesus; and when they saw him they begged him to leave the district and go./85/

In the revision of this story by Mark and Luke, there is only one demoniac. Jesus asks him, "What is your name?"; to which the demoniac responds "my name is Legion." "This was because," says Luke, "so many devils had taken possession of him. And they begged him [Jesus] not to banish them to the Abyss."/86/

Kierkegaard-Haufniensis observes that the phenomenon of the demonic has previously been interpreted in three ways: (1) *as an aesthetic-metaphysical phenomenon* where the demonic comes under the rubrics of misfortune, fate, etc., and is treated with sympathy that one would treat, for example, a madness resulting from a birth defect; (2) *as an ethical phenomenon* where the demoniac has been condemned, persecuted, discovered, and punished; or finally (3) *as a medical-therapeutical phenomenon*, where the demoniac is treated as a sufferer who requires treatment in order to be cured.

These three interpretations correspond to the somatic, psychic, and pneumatic spheres of being human, a correspondence suggesting that the phenomenon of the demonic is of far greater scope and complexity than had previously been supposed. Indeed, Kierkegaard-Haufniensis says: ". . . there are traces of it in every man, as surely as he is a sinner" (*CA*, p. 122). It is necessary, therefore, to examine the demonic experimentally and psychologically in order to show that Angst is at its epicenter. Says Kierkegaard-Haufniensis:

> The demonic is anxiety [Angst] about the good. In innocence, freedom was not posited as freedom; its possibility was anxiety in the individual. In the demonic, the relation is reversed. Freedom is posited as unfreedom, because freedom is lost. Here again freedom's possibility is anxiety. The difference is absolute, because freedom's possibility appears here in relation to unfreedom, which is the very opposite of innocence, which is a qualification disposed toward freedom. (*CA*, p. 123)

On the strength of this observation, Kierkegaard-Haufniensis concludes: "the demonic is *inclosing reserve, and the unfreely disclosed*" (*CA*, p. 123). Thus we see that the demonic's characteristic, inclosing reserve, is precisely that which Kierkegaard saw in his father, as well as that which drove the young Kierkegaard to develop the "master-thief" concept. Specifically, inclosing reserve is the mute, and self-expression of what is mute must "take place contrary to its will, since freedom . . . revolts and now betrays unfreedom in such a way that it is the individual who in anxiety [Angst] betrays himself against his will" (*CA*, p. 123).

In a haunting passage that cannot help but be autobiographical, Kierkegaard-Haufniensis betrays his veil of pseudonymity; and with the ghost of Michael Kierkegaard before him, he confesses: "It is incredible what power the man of inclosing reserve can exercise over such people, how at last they beg and plead for just a word to break the silence, but it is also shameful to trample upon the weak in this manner" (*CA*, p. 125). Thus the psychological method is used to ferret out the enormous power of the demonic. The good, against which the inclosing reserve of the demonic rebels, is manifested in the act of "disclosure" as the means whereby the dark secret of the demoniac may be released. For what determines whether a phenomenon is demonic, says Kierkegaard-Haufniensis, is ". . . the individual's attitude toward disclosure, whether he will interpenetrate that fact with freedom and accept it in freedom. Whenever he will not do this, the phenomenon is demonic" (*CA*, p. 129).

Hence, inclosing reserve and its relationship to disclosure is a theme grounded in Kierkegaard's experimental/psychological method. This approach was used in attempting to get his father to disclose his dreadful secret which in Kierkegaard's view would restore Michael Kierkegaard to freedom. Kierkegaard-Haufniensis now introduces into these psychological/experimental considerations the principle of consistent thought. As a result, the rest of *Begrebnet Angest*'s Chapter IV demonstrates the breadth of his cognitive method by consistently thinking through the implications of the demonic. The chapter culminates in a graphic discussion of the essential attributes or characteristics of *inwardness* and *certitude* as what has *principally been lost* in the demoniac's commitment to unfreedom. It may be worthwhile here to explore these characteristics in brief, for they will re emerge as important dimensions of Heidegger's conception of Angst in the next chapter.

For Kierkegaard-Haufniensis the definition of "inwardness" is given as "earnestness." Earnestness, in turn, is likened to Rosenkranz's definition of "disposition" [*Gemyt*]./87/ Since we have used the term "pre-disposition" to characterize the *Befindlichkeit* of Angst in our previous chapter, here an examination of how "disposition" is used respectively by Rosenkranz and Kierkegaard-Haufniensis may prove useful. For Rosenkranz, then, the term "disposition" refers to a specific psychological state within the individual, experienced as a unity of feeling and self-consciousness./88/ He explains: "The feeling unfolds itself into self-consciousness and vice versa, that the content of self-consciousness is felt by the subjects as his own. It is only this unity that can be called disposition."/89/

From the perspective of Kierkegaard-Haufniensis, the concept of earnestness "is a higher as well as the deepest expression for what disposition is" (*CA*, p. 148). Rather than becoming a dogged, habitual way of conducting one's existence, therefore, earnestness can never become mere habit, because it is constantly a fresh and original relationship to one's existence; that is, earnestness is a relationship to personal existence qualified by the eternal rather than the temporal. "Whenever inwardness is lacking, the spirit is finitized. Inwardness is therefore eternity or the constituent of the eternal in man" (*CA*, p. 151).

With these observations we leave the phenomenon of the demonic as the paradigm of the experimental/psychological approach to Angst. As Kierkegaard-Haufniensis himself was fond of saying in another context, "further than this psychology cannot go" (*CA*, p. 45). Thus at bottom, the demonic is a state where freedom is lost, where the truth of inwardness and certainty is lost, and where the demoniac is inclosed in a melancholic reserve. The demoniac is thus deprived of faith in the eternal of Christianity—the only means for allaying, but never eliminating, the Angst that is a necessary part of his human existence.

In the closing chapter of the work, Kierkegaard-Haufniensis offers a summary description of Angst.

> Anxiety [Angst] is freedom's possibility, and only such anxiety is through faith absolutely educative, because it consumes all finite ends and discovers all their deceptiveness. And no grand inquisitor has such dreadful torments in readiness as anxiety has, and no secret agent knows as cunningly as anxiety how to attack his suspect in his weakest moment or to make alluring the trap in which he will be caught, and no discerning judge understands how to interrogate and examine the accused as does anxiety, which never lets the accused escape, neither through amusement, nor by noise, nor during work, neither by day nor by night. (*CA*, pp. 155–56)

From this passage Kierkegaard-Haufniensis concludes: to be tutored by Angst is to be *educated by the infinity of possibility*. But such an education demands from the individual a painful tuition. When one faces possibilities as *ontological* possibility while grasping at his own finitude before the abyss, one realizes with equally profound ontological Angst that "he can demand nothing from life and that the terrible, perdition, and annihilation live next door to every man . . ." (*CA*, p. 156). Thus the lesson to be learned in this school is the burden of facing one's possibilities with passionate inwardness. The ability to attain this end comes only through faith, understood by Kierkegaard-Haufniensis as "the inner certainty that anticipates infinity."/90/ Faith, then, is the only means for alleviating the daunting ontological Angst which the individual must accept in the school of possibility. Says Kierkegaard-Haufniensis:

> He who sank in possibility . . . sank absolutely but then in turn he emerged from the depths of the abyss lighter than all the troublesome and terrible things in life. . . . For him, anxiety [Angst] becomes a serving spirit that against its will leads him where he wishes to go. . . . Then anxiety enters into his soul and searches out everything and anxiously torments everything finite and petty out of him, and then it leads him where he wants to go. (*CA*, pp. 158–59)

Thus only after having attained the standpoint of faith can Angst take on an entirely new significance for the human spirit; Angst now serves as a potent purgative power whereby the petty, the superficial, and the finite are banished from the spirit, allowing that spirit to boldly dwell in faith's eternal promise. Further than this Kierkegaard-Haufniensis as psychologist cannot go. To do so he must take up the gauntlet of the Christian writings, a challenge clearly beyond the pseudonymous author of *Begrebnet Angest*. So Kierkegaard-Haufniensis concludes his work with, "Here this deliberation ends where it began. As soon as psychology has finished with anxiety [Angst], it is delivered over to dogmatics" (*CA*, p. 162).

Preliminary Conclusions

To be sure, we have touched upon a large amount of material in this chapter in the need to rigorously describe the horizons of Kierkegaard's aesthetic thought on our way to the phenomenon of Angst. As to the *adequacy* of Kierkegaard's ontical depiction of Angst, the chapter following will display how Kierkegaard's description of Angst's essential structures provides the ontic foundation for Heidegger's ontological and phenomenological interpretation of Angst. Indeed we may state that virtually every ontic feature of Heidegger's Angst description and interpretation is grounded in Kierkegaard's *Begrebnet Angest*.

On the other hand, it seems to us that a hermeneutic phenomenological disclosure of Angst's full ontological meaning was not possible for Kierkegaard. This has nothing to do with any lack of rigor or cognitive acuity on Kierkegaard's part. Nor has it to do specifically with the hereditary sin context in which

Kierkegaard reveals Angst's meaning. We believe, rather, that Kierkegaard's inability lies precisely in the historical parameters of philosophy at the time Kierkegaard wrote. More specifically, Kierkegaard was prevented from revealing the full hermeneutical significance of Angst by his entrapment within the assumptions of Western philosophy at the time, assumptions that were *so* ingrained in the intellectual tradition of the West that they were not challenged until Nietzsche. There are two basic reasons for this entrapment.

First, while his principal concern was to reveal the ethical dimension of human existence, what Blackham/91/ has seen as the "how" of existential choice, it remains clear that Kierkegaard uncritically accepted the philosophical view that at root existential man remains a *res cogitans*, a "thinking substance," to which the predicates of understanding, feeling, and will are attached. While Kierkegaard takes despair as his point of departure rather than Cartesian doubt, man is conceived as subjective spirit rather than pure Dasein./92/ Without rehashing how this leads inevitably through "subject" to ontological solipsism, we remark only in passing that the consequences of Kierkegaard's acceptance of man as "res cogitans" casts grave reservations on the adequacy of the *unum noris omnes* principle to account for the existence of the "existential categories" or qualifications of the human spirit *in others*.

Secondly, Kierkegaard's reliance on the rigorous logic of consistent thinking to refine the "know thyself!"; his further reliance on the *unum noris omnes* to universalize these findings; and finally, his reliance on experimentation and psychological observation to confirm these findings *all reveal* his unexamined dependence on the principles of the human sciences to verify, confirm, and substantiate something like "objective truth," even in the face of the subjective truth which for Kierkegaard was the highest truth for the existing individual. Thus while ethical or existential truth is characterized as "objective *uncertainty* held fast in an appropriation process of the most passionate inwardness,"/93/ the truths of anthropological contemplation rely on objective *certainty* attained only in a verification process of passionately interested observation. To entertain both of these truth models simultaneously may have been possible for Kierkegaard's dialectical genius, but as we shall see in the next chapter, the first of these views leads down the path to psychologism, and the second naively assumes that the principles of natural science can successfully be applied universally to the human sciences. As we shall see presently, the birth of Husserl's phenomenology is specifically traced to his repudiation of psychologism as a vacuous enterprise, and Dilthey's "philosophy of life" (*Lebensphilosophy*) finds its strength in showing that the principles of *natural* science specifically *do not* apply to the human sciences.

In fairness to Kierkegaard, however, we are compelled to state that all of these criticisms are appropriate only with regard to his purely aesthetic and pseudonymous writings. After all, Kierkegaard's primary concern was *not* with positivistic science, but rather with the canons of the subjective thinker, the thinker whose *existence* is the decisive issue rather than his objective essence. Thus we are not claiming that Kierkegaard was a closet positivist, but merely that the assumptions of the human sciences permeated the intellectual climate of opinion to which Kierkegaard belonged. Indeed, it is precisely against such assumptions that much of Kierkegaard's polemic writings are directed, especially those which take exception with Hegel.

Turning from these broad concerns to Angst in Kierkegaard's thought specifically, we may conclude, indeed, Kierkegaard did set the stage for Angst to reveal its role in later philosophical thought. Kierkegaard's discussions of the distinction between Angst and fear, for example, or his interpretations of nothingness, freedom, temporality, and resoluteness, for example, established a permanent *ontic* foundation for later phenomenological and existential revelations of Angst. Yet, we must likewise observe that Kierkegaard's analysis does not, indeed, *cannot*, get beyond the ontic "vulgar"/94/ psychological interpretation which covers over the full ontological role so thoroughly described and interpreted by Heidegger.

NOTES

/1/ Matthew 12:31 (NEB).

/2/ The term "existentialism" is in our view a complete misnomer. To exist means in some sense to "stand-out" from the background, to be unique; a point which Heidegger, for one, insists upon. Thus there cannot be an "ism" of existence since -ism implies a coherent system of doctrine. If each Dasein ex-ists in the sense of standing out individually, then there can be no collective doctrine or -ism of such existence. The two are incommensurate. Cf. note 66 below.

/3/ Rollo May et al., *Existence: A New Dimension in Psychiatry and Psychology* (New York: Simon and Schuster, 1958), Chapter 1.

/4/ Howard V. Hong and Edna H. Hong, "Translator's Foreword," in Gregor Malantschuk, *Kierkegaard's Thought*, ed. and trans. by Howard V. Hong and Edna H. Hong (Princeton: Princeton University Press, 1971), p. vii.

/5/ Kierkegaard's *Journals and Papers* demonstrate an in-depth understanding of the Church Fathers. Cf. Howard V. Hong and Edna H. Hong, editors and translators, *Søren Kierkegaard's Journals and Papers*, 7 vols. (Bloomington and London: Indiana University Press, 1967–78) (hereafter cited as *JP*). We shall cite the volume number plus the Hong's entry number and the number in the *Journals and Papers* assigned by the Danish editors of *Søren Kierkegaard's Papirer*, 20 vols., I-XI (Copenhagen: Gyldendal, 1909–49). In the latter series of numbers, "A" designations refer to Journal entries, "B" to drafts of published works, and " C" to reading notes. Then, if appropriate, we shall provide the date if given that the entry was written. Thus for the references to the Church Fathers, see *JP*, vol. 1, 583 (II A 750); vol. 2, 2867 (X A 288); vol. 4, 38:30 (X A 119); vol. 4, 4093 (II A 436); and vol. 6, 6677 (X A 434).

/6/ *JP*, vol. 1, 219 (II A 127) and 1309 (237); vol. 5, 5227 (599) and 5350 (II A 702). Also see *Manichaeism* entry in vol. 1, 1302 (I A 2).

/7/ *JP*, vol. l, 884 (III A 125); vol. 5, 5010 (VIII A 105).

/8/ *JP*, vol. 6, 6397 (X A 323).

/9/ Franz X. von Baader, *Vorlesungen . . . über religiouse Philosophie* (Munich 1927; *ASKB* 395); *Sämmtliche Werke*, I-XVI (Leipzig, 1850–60).

/10/ There is no mention of Eckhart, as far as we know, in any of Kierkegaard's works.

/11/ *JP*, vol. 1, 96 (III A 235), n.d., 1842.

/12/ Søren Kierkegaard, *The Concept of Anxiety*, translated by Reidar Thomte (Princeton: Princeton University Press, 1980), p. 162, footnote (hereafter when a passage is cited directly from this translation, we will follow the convention of referring to the source as "*CA*," and then give the appropriate page number). Cf. note 71 below.

/13/ *JP*, vol. 1, 94 (III A 233), n.d., 1842.

/14/ *Papirer*, V B 53:18, n.d., 1844; as cited in *CA*, "Entries from Kierkegaard's Journals and Papers," p. 187.

/15/ Kierkegaard sided with Augustine against Pelagius concerning the doctrine of hereditary or original sin on the grounds that the Pelagian view ". . . permits every individual to play his little history in his own private theater unconcerned about the race"; cf. *CA*, p. 34.

/16/ Vincent A. McCarthy, *The Phenomenology of Moods in Kierkegaard* (The Hague/Boston: Martinus Nijhoff, 1978), p. 47.

/17/ Ibid.

/18/ Ibid.

/19/ *Papirer*, I, pp. xi-xvi, as cited in Gregor Malantschuk, *Kierkegaard's Thought*, p. 11. Note: we are indebted to Malantschuk's discussion of Kierkegaard's dialectical method as the source of our own view of the two approaches discussed in this section. It is generally agreed that Malantschuk's contribution is exceedingly important in understanding Kierkegaard's thought. Thus we have no reservations in relying on the hermeneutical soundness of his interpretations, which like our own seek confirmation across the broad spectrum of Kierkegaard's authorship. We are, therefore, extremely grateful to Malantschuk and wish to acknowledge that fact here.

/20/ Malantschuk, *Kierkegaard's Thought*, p. 11.

/21/ *JP*, vol. 5, 5100 (I A 75), August 1, 1835.

/22/ Ibid.

/23/ Malantschuk, *Kierkegaard's Thought*, p. 109.

/24/ *JP*, vol. 5, 5100 (I A 75), August 1, 1835.

/25/ Søren Kierkegaard, *Concluding Unscientific Postscript*, trans. by David F. Swenson and Walter Lowrie (Princeton University Press, 1941), p. 318, as cited in Kresten Nordentoft, *Kierkegaard's Psychology*, trans. by Bruce H. Kirmmse (Pittsburgh: Duquesne University Press, 1978), p. 24.

/26/ *JP*, vol. 3, 3675 (IVC78), n.d., 1842–43.

/27/ *Papirer*, V B 53, p. 112.

/28/ *Papirer*, V B 49, p. 110; note: this entire indirect proof is constructed by Nordentoft rather than by us, and it is cited in Nordentoft, *Kierkegaard's Psychology*, p. 6.

/29/ Malantschuk, *Kierkegaard's Thought*, p. 112.

/30/ Ibid., p. 105.

/31/ Ibid., p. 106.

/32/ *CA*, pp. 25–26, 28, 32–36, 122, 207.

/33/ F. C. Sibber, *Om Erkjendelse og Granskning* (Copenhagen, 1822); as cited in Malantschuk, *Kierkegaard's Thought*, p. 107.

/34/ Malantschuk, *Kierkegaard's Thought*, p. 108.

/35/ F. C. Sibbern, *Bemaerkninger og Undersogelser, fornemmelig betraeffende Hegels Philosophi, betraget i Forthold til vor Tid* (Copenhagen, 1838) (hereafter *Bemaerkninger*); as cited in Malantschuk, *Kierkegaard's Thought*, p. 126.

/36/ Ibid. Note: the quote is from Malantschuk, not Sibbern.

/37/ We had best deal with this term at this point. "Qualifications of the mental-spiritual" in human existence are existential categories of being human, much like the *Existenzials* of Heidegger's thought. Thus they are *not* categories in the sense of Aristotle or Kant, because categories apply to things, not humans.

/38/ Sibbern, *Bemaerkninger*, p. 130; as cited in Malantschuk, *Kierkegaard's Thought*, p. 128.

/39/ Sibbern, *Bemaerkninger*, p. 131; as cited in Malantschuk, *Kierkegaard's Thought*, p. 128.

/40/ Malantschuk, *Kierkegaard's Thought*, p. 128.

/41/ Ibid., pp. 129–30.

/42/ *JP*, vol. 3, 3276 (II A 517), July 28, 1839.

/43/ Malantschuk, *Kierkegaard's Thought*, p. 257.

/44/ Nordentoft, *Kierkegaard's Psychology*, pp. 2–3.

/45/ *Papirer*, V B 53, p. 112; as cited in Nordentoft, *Kierkegaard's Psychology*, p. 6.

/46/ Ibid.

/47/ This is, of course, the view of the pseudononymous author Victor Ememita, the "victorious" religious recluse or solitary individual of *Either/Or*. Yet, the auditory method is confirmed by Virgilius Haufniensis in *Begrebnet Angest*, so there is reason to suspect that it grew out of Kierkegaard's own approach to psychology. We shall consider this point again when we discuss the "master-thief" idea in connection with the phenomenon of "presentiment."

/48/ *Papirer*, V B 53, p. 119; as cited in Nordentoft, *Kierkegaard's Psychology*, p. xvii.

/49/ Nordentoft, *Kierkegaard's Psychology*, p. 4; note: Nordentoft does not identify the source of this quote.

/50/ Ibid., p. 5.

/51/ *JP*, vol. 5, 5664 (IV A 107), May 17, 1843.

/52/ Ibid.

/53/ This difficult term, like the term "Angst," has no exact equivalent in English. It has been translated by Thomte as "inclosing reserve," by Lowrie as "shut-upness," a word that is certainly more graphic than "inclosing reserve," and by Malantschuk as "closed-up-ness." Thomte's translation is employed here because "inclosing reserve" seems to have the flavor of a basic disposition, again in a Shakespearian sense, as a mode of being toward the demonic. This will become clear in Section II 3. Thomte's translation is in *CA*, p. 123; Lowrie's in *CD*, p. 110, and Malantschuk's in Malantschuk, *Kierkegaard's Thought*, p. 27.

/54/ *JP*, vol. 5, 5074 (I A 15), January 29, 1835.

/55/ *JP*, vol. 1, 91 (II A 18), n.d., 1837.

/56/ Ibid.

/57/ *CA*, pp. 61–62.

/58/ *JP*, vol. 3, 3551 (II A 32), n.d., 1837.

/59/ *JP*, vol. 3, 3553 (II A 558), September 10, 1835.

/60/ *JP*, vol. 3, Notes, p. 886.

/61/ Søren Kierkegaard, *Repetition*, trans. by Walter Lowrie (Princeton: Princeton University Press, 1941), as cited in *JP*, vol. 3, Notes, p. 886.

/62/ *JP*, vol. 4, 3999 (II 584), n.d., 1837.

/63/ *JP*, vol. 1, 94 (III A 233), n.d., 1842.

/64/ We are not suggesting, of course, that Kierkegaard himself was explicitly aware of the difference between the ontic and the ontological as was Heidegger later, or even that Angst was understood and interpreted by Kierkegaard himself as Dasein's *Grundefindlichkeit*. But we *are* suggesting that all of these notions are present but unarticulated in Kierkegaard's thought.

/65/ *JP*, vol. 1, 100 (VIII A 363), n.d., 1837.

/66/ *CA*, p. 57. Note: In an extremely important footnote, Thomte clarifies the meaning of *Tilvaerelse*. It is worth citing in its entirety. He notes: "The Danish terms *Tilvaerelse* (vb. *vaere til*) and *Eksistens* (vb. *eksistere*) are both translated into English by the word 'existence' (vb. to exist). *Tilvaerelse* corresponds to the German word Dasein (*was da ist*), and it usually denotes the outer observable existence in time and space. To make more explicit the distinction between 'existence' in the existential sense and 'existence' as an outward observable existence, the German word Dasein might well be used for the latter" (*CA*, "translator's notes," p. 226, note 26.) This, it almost goes without saying, is not the meaning Heidegger gives to Dasein, and if Thomte is correct, then Heidegger has his terms for existence reversed.

/67/ *JP*, vol. 5, 5383 (II A 420), May 12, 1839.

/68/ Nordentoft, *Kierkegaard's Psychology*, pp. 13–15.

/69/ The draft of the title page has Kierkegaard's name as author. Cf. *Papirer*, V B 42, n.d., 1844.

/70/ Since it is difficult to determine precisely where Haufniensis differs from Kierkegaard, we will follow McCarthy's example of using both names as individuals responsible for the work. Cf. McCarthy, *Moods in Kierkegaard*, p. 36.

/71/ Quotes directly from the *Concept of Anxiety* will follow the passage cited without footnotes unless one is needed. This will provide quicker access to the reference for the reader.

/72/ Kierkegaard, in his categorization of the *Geisteswissenschaften*, included both dogmatics and psychology under the subjective rubric. Cf. Malantschuk, *Kierkegaard's Thought*, p. 139.

/73/ The reader will recall that psychology as a discipline had in Kierkegaard's time a much broader meaning than it does today.

/74/ Supra, Chapter I.

/75/ As Thomte points out, there is much debate over this intitial characterization from within the corpus of Kierkegaard's own drafts. The Danish editors of Kierkegaard's *Samlede Vaerker*, says Thomte, "assume that the term 'the possibility of possibility' is a slip of the pen and that the intended reading is 'the possibility of freedom.'" Thomte does not think that this is tenable. He argues that the term "possibility" must be understood in relation to its context; namely, "When man is psychically qualified in unity with his naturalness, and the spirit is *sleeping*, human freedom does not manifest itself. Anxiety [*sic*] is the qualification of the dreaming spirit, and when the spirit becomes *awake*, the difference between oneself and the other is posited. In the dreaming state, spirit has a presentiment [!] of the freedom that follows when consciousness is awakened. This presentiment of freedom, this state of anxiety, is spoken as 'freedom's actuality as the possibility of possibility.' It is also spoken of as 'entangled, not by necessity, but in itself.' Cf. 'translator's notes,' p. 235, note 46."

/76/ *JP*, vol. 1, 94 (III A 233), n.d., 1842.

/77/ Walter Kaufmann, *Discovering the Mind*, vol. II (New York: McGraw Hill, 1980), p. 25.

/78/ McCarthy, *Moods in Kierkegaard*, p. 41.

/79/ Supra, p. 49.

/80/ Thomte points out that the Danish word is *Øiblikket* [the moment], comparable to the German, *Augenblick*, meaning "a blink of the eye." In *CD*, Lowrie translated *Øiblikket* 5 as "the instant," but this loses the sense of continuity with the Latin *momentum* meaning from *movere* [to move] the merely vanishing. Compare *CA*, pp. 87–88, with *CD*, pp. 78–79.

/81/ *CA*, "translator's notes," p. 245.

/82/ Supra, note 53.

/83/ The discussion concerning Angst about evil runs approximately four pages in *Begrebnet Angest* as opposed to the thirty-six or so pages concerning Angst about the Good and its attendant loss of freedom. It is reasonable to conclude, therefore, that Angst of evil, much like the section on objective Angst discussed above, may be seen as an architectonic device that allows Kierkegaard-Haufniensis to contrast the demonic with repentance. The major point of this discussion is that repentance is virtually a useless passion in the face of Angst, since it cannot free the individual from his Angst. Only faith can accomplish that, and the discussion of faith is reserved for the concluding chapter of *Begrebnet Angest*.

/84/ Kierkegaard-Haufniensis defines the good in the following ironic way: "The good cannot be defined at all. The good is freedom. The difference between good and evil is only for freedom and in freedom, and this difference is never *in abstracto* but *in concreto*" (*CA*, p. 111).

/85/ Matthew 8:28–34 (NEB).

/86/ Luke 8:30–31 (NEB).

/87/ Karl Rosenkranz, *Psychologie oder die Wissenschaft vom Subjecktiven Geist* (Koenigsberg, 1837), p. 322; as cited in *CA*, p. 148.

/88/ Ibid.

/89/ Ibid.

/90/ Kierkegaard-Haufniensis paraphrases this definition, says Thomte, from either of two passages from Hegel: (1) "Faith must be defined as the witness of the spirit to absolute spirit, or as a certainty of the truth," or (2) "Faith may be defined as being a witness of the spirit to spirit, and this implies that no finite content has any place in it"; cf. G. W. F. Hegel, *Philosophie der Religion*, Part One, C, I, 2, *Werke*, XI, pp. 206, 213; *J.A.*, XV, pp. 222, 229; *Lectures on the Philosophy of Religion*, I, pp. 212, 218; as cited in *CA*, "translator's notes," p. 253.

/91/ H. J. Blackham, *Six Existentialist Thinkers* (New York and Evanston: Harper Torchbooks, 1959), pp. 15–16.

/92/ Kierkegaard, *Either/Or*, Vol. II, p. 178.

/93/ Kierkegaard, *Concluding Unscientific Postscript*, p. 182.

/94/ By "vulgar" we do not mean "tasteless." Rather, we use the term as in the Latin *vulgarus* meaning "common" as distinguished from "refined" or, more appropriately perhaps in this case, "esoteric."

CHAPTER III

HEIDEGGER AND ANGST

> In the abyss of terror, courage recognizes the virtually unexplored realm of Being: that openness into which each entity returns as what it is and what it can be.
>
> —Martin Heidegger, "Postscript" to *What is Metaphysics?*

The Question at Hand

On a crisp wintry morning in the early 1920s, a young German scholar places another log on the fireplace. As the warmth fills the room, he squints through his window, drinking in the freshness of the Black Forest snowscape. Just outside the window a battered wooden water pump stands half-buried in the snow./1/ Beyond the pump and down the hill, an evergreen tree, its branches laden with snow, obscures the scholar's vision of a skier, slithering through the moguls of the powdered slopes.

The scholar turns to the blank page before him. He picks up his nib-point pen and dips it into the inkwell. With a moment's glance at the portrait of Blaise Pascal hanging on the wall,/2/ he bends to his task, penning: "40. *Die Grundbefindlichkeit der Angst als eine ausgezeichnete Erschlossenheit des Daseins*" [The primordial disposition of Angst as a distinctive means of disclosing Dasein]./3/ Another morning's work begins for the young Marburg philosophy professor, Martin Heidegger.

With this vision in mind we may state the purpose of this chapter. Very broadly speaking, we have to come to grips with what happened that morning in Heidegger's Black Forest study, by understanding his notion of primordial Angst and its role in the early mature period of his thought, the period from 1921–1929./4/ During these years Heidegger spent considerable time working through the implications of primordial Angst, specifically addressing the phenomenon in four major works: *Being and Time*,

What is Metaphysics?,/5/ *On the Nature of Ground*,/6/ and *Kant and the Problem of Metaphysics*./7/ Of these four, only the first two will concern us since they represent the principal works in which Angst plays a significant role. While primordial Angst is mentioned, in the later two works, Heidegger introduces no further substantive material than what went on before in *Being and Time* and *What is Metaphysics*?

Our main task here, therefore, is to grasp essentially Heidegger's Angst interpretation in *Being and Time* and in *What is Metaphysics*? We believe the evidence shows that Heidegger developed Kierkegaard's psychological-ontic, theological Angst into a full-dress ontico-ontological description of *primordial Angst* as we ourselves discussed that term in Chapter I.

The role of primordial Angst in Heidegger's early mature period has been misunderstood by many, abused by some, but *essentially* grasped by none save Heidegger—all of which may account for why he abandoned primordial Angst as an explicit theme for further investigation after 1929./8/ Yet, Heidegger's Angst interpretation, like Kierkegaard's before him, serves as a bridge to later interpretations. Hence we do not claim that Heidegger's is the final or even the furthest-reaching Angst interpretation: Rather his is but another major milestone along the path to fully grasping Angst's essential meaning.

Thus to prepare the hermeneutic background for understanding Angst from Heidegger's perspective we must very carefully carve out the horizons that make that interpretation possible. And, following the micro-hermeneutical precedent of the previous chapter, this means moving from the broadest to the most narrow focus of Heidegger's thought concerning primordial Angst. As in the previous chapter, our task is to build the framework, to set the lighting, so to speak, so that the rich colors of Heidegger's primordial Angst portrait may be revealed in their rich but subtle hues. For primordial Angst is not merely some muted background theme in Heidegger's philosophy; rather Angst is a vivid centerpiece that reveals the connection point between Dasein and world in *Being and Time*. This centerpiece becomes highlighted in *What is Metaphysics?* as a more explicit connector between Dasein and scientific understanding in general.

The Horizons of Influence

One and only one philosophical question guided Heidegger through the forest-path of his philosophical thought: the omnipresent "*Seinsfrage*"/9/—the question of the meaning of Being. We shall see how Heidegger came to the "Being-question" momentarily, but for the present let us examine this haunting question's initial articulation. Heidegger begins *Sein und Zeit* with a passage from Plato's *Sophist* (244A), where the stranger says to Theaetetus: "Obviously you must be quite familiar with what you mean when you use the expression 'being,' whereas we, who formerly imagined we knew are now perplexed." Heidegger's comments on this passage are his point of departure for *Sein und Zeit*. He remarks:

> In our own time do we have an answer to the question of what we mean when we use the expression "being." Absolutely not. It is fitting and proper therefore that we reawaken the question of the meaning of Being. But are we even perplexed or embarrassed by our lack of understanding "Being"? Not at all! So first and foremost it is necessary to awaken anew an appreciation of the very meaning of this question. The aim of the following treatise is to work out concretely the question of the meaning of Being. (*SZ*, p. 1)

To have access to Heidegger's general thought, to *understand* Heidegger at all, we must recognize that the question of Being permeates Heidegger's thinking. Thus Heidegger's analytic of Dasein, his illumination of Dasein's ontological categories, his discussion of primordial Angst and its relationship to death, and his descriptions of authenticity and temporality are all undertaken in the name of the omnipresent question of Being. How, then, did Heidegger come to this initial question? To answer this question we must explore the influences on Heidegger's thought prior to *Being and Time*.

The clearest expression of both Heidegger's philosophical training and his philosophical apprenticeship at Marburg is provided in Heidegger's brief but biographically pregnant essay, "My Way to Phenomenology."/10/ In this work, Heidegger informs us that he was committed to theology rather than philosophy during his early years at Freiburg. Yet, during his spare time, he was able to read with growing enthusiasm Husserl's *Logical Investigations*. Heidegger explains why he spent his free time in this manner.

> I had learned from many references in philosophical periodicals that Husserl's thought was determined by Franz Brentano. Ever since 1907, Brentano's dissertation "On the Manifold Meaning of Being since Aristotle" (1862) had been my chief help and guide of my first awkward attempt to penetrate into philosophy./11/

Franz Brentano, then, was Heidegger's first philosophical mentor rather than Husserl. Brentano's philosophical problematic, "the Manifold Meaning of Being," remained with Heidegger long after he had abandoned Husserl's transcendental or "pure" phenomenology. But at the outset, Heidegger absorbed Husserl's *Logical Investigations* with only one end in mind: to gain an answer to the question, ". . . what is its [Being's] fundamental meaning/What does Being mean?"/12/ This first plunge into Husserl's thought failed miserably, Heidegger admits; ". . . my efforts" says Heidegger, "were in vain because I was not searching in the right way."/13/

By 1911, two years after entering Freiburg as a student, Heidegger's enchantment with the question of Being burgeoned to a point where he abandoned formal theology. But while he was now committed to the study of philosophy, the humanities, and science,/14/ he still spent considerable time discussing with the theologian Carl Braig "Schellings's and Hegel's significance for speculative theology as distinguished from the dogmatic system of scholasticism."/15/ Probably Heidegger's first encounter with primordial Angst came through these discussions, and specifically in Braig's comments concerning Schelling's interpretation of Jacob Boehme. While no direct evidence exists to prove this hypothesis conclusively, we offer it with the stipulation that the Braig-Schelling-Boehme Angst connection remains a good possibility, one that makes sense./16/ Nonetheless, the fact that Heidegger maintained an active interest in theology, even after devoting himself to philosophy, shows he still considered theology to be a viable alternative to ontology as the proper structure of metaphysical inquiry./17/

At this time Heidegger was led once again to explore Husserl's phenomenological thought through the medium of Emil Lask, whose two works, *The Logic of Philosophy and the Doctrine of the Categories, A Study of the Dominant Realm of Logical Form* (1911) and *The Doctrine of Judgement* (1912), showed the influence of Husserl's *Logical Investigations*. Heidegger was now approaching phenomenology./18/ Once again, however, Heidegger was frustrated in his attempt to understand Husserl; he could not fully grasp "how thinking's manner of procedure which called itself 'phenomenology' was to be carried out."/19/ The problem, says Heidegger, was that he was virtually transfixed by an essential ambiguity in Husserl's work. Specifically, in Volume I of the *Logical Investigations* Husserl takes great pains to refute "psychologism" in logic by showing that thought and knowledge in principle are not reducible to pure psychology. But in Volume II of the same work, Husserl describes "psychological" acts of consciousness as being necessary for the constitution of knowledge. Clearly a contradiction exists between Volume I and II regarding the status of psychologism. "Accordingly," Heidegger concludes, "Husserl falls back with his phenomenological description of the phenomena of consciousness into a position of psychologism which he had just refuted."/20/

There was yet a greater problem. Husserl's later insistence in *Ideas I* that pure phenomenology ". . . as the most fundamental region of philosophy, is an essentially new science . . ."/21/ served to alienate Heidegger further. Heidegger believed that Husserl's new vision placed philosophy firmly under the shadow of science, since it made science the paradigm for any proper philosophical investigation. Heidegger simply could not accept such a position, and he therefore rejected transcendental or pure phenomenology at the outset of his philosophical career. This explains why the terms "consciousness," "idea," "ego," etc., are not a general part of Heidegger's philosophical lexicon.

In spite of these differences, the rift between Heidegger and Husserl had not yet reached he point where reconciliation was impossible. Indeed, Husserl remained Heidegger's mentor, because Heidegger believed Husserl's Sixth Logical Investigation provided a more meaningful basis for the enterprise of phenomenology. Specifically, Heidegger saw in Husserl's Sixth Investigation that the phenomenological notion of the self-manifestation of phenomena had been conceived more originally by the Greeks, Aristotle in particular, as *aletheia*, or the unconcealedness of what is present in self-revelation./22/ Heidegger believed this unconcealing, this self-revealing of and by the phenomena themselves, was an absolutely fundamental breakthrough for phenomenology. "That which phenomenological investigation rediscovered as the supporting attitude of thought,"

Heidegger comments, "proves to be the fundamental trait of Greek thinking, if not indeed of philosophy as such."/23/

This earth-shattering insight into *aletheia* as a more primordial phenomenology unlocked a new door for Heidegger. Beyond this door stood a new horizon of phenomenological investigation, one centered upon seeing phenomenology as discipline rooted in the history of the Western ontological tradition. Accordingly, Western ontology must be called into account to deal adequately with the historical dimension of the question of Being. Nonetheless, the truth of *aletheia* and the fact of its utter neglect throughout the history of the Western metaphysical tradition, displays graphically the enormous power of Western philosophy to cover over and deeply bury fundamental insights into the question of the meaning of Being. Thus history must be considered along with phenomenology to yield the full fruit of philosophical ontology. Husserl's vision manifestly failed to grasp this significant dimension, according to Heidegger's reckoning.

Heidegger found the means to overcome this lack by absorbing the *Lebens-philosophie* (philosophy of life) of Wilhelm Dilthey and Dilthey's friend, Count Paul von Yorck. Together these two scholars constitute the third major influence on Heidegger during his early mature period. Specifically, Dilthey sought to ground philosophy in the experience of life itself, an experience he called "lived-experience" (*Erlebnisse*), rather than in the cognitive apodicticity of consciousness à la Husserl. By "lived-experience" Dilthey meant ". . . teleological units of meaning . . . objectified in the cultural productions of man . . . connected together into the structural unity of his life history."/24/ This approach, in contrast to Husserl's demand for cognitive apodicticity, emphasizes the meaning of being human as the primary reality that must precede any understanding of the things themselves. The meaning of being human is necessarily historical, therefore, since lived-experiences are made meaning-*full* within the context of our historical, cultural reality. Accordingly, Dilthey observes: ". . . man does not understand his own self by means of any kind of rumination upon himself . . . only through an understanding of the historical reality generated by him does he obtain a consciousness of his capacities, for good or for ill."/25/ Thus to obtain access to self-understanding as the alpha point of philosophical investigation requires approaching human existence from the standpoint of the human sciences (*Geisteswissenschaften*). In these human sciences apodicticity is not the result of the causal explanations of natural science's mathematical paradigms. Rather apodicticity results from placing human understanding (*Verstehen*) ahead of scientific explanation (*Erklaerung*). For Dilthey, the appropriate methodology for obtaining self understanding was hermeneutics, the art of interpreting life itself.

While Heidegger found a more satisfactory approach to the problem of the question of Being (seen in terms of human being) in Dilthey's thought, there remained in Dilthey's philosophy of life a basic and fatal weakness: an unexamined presupposition that takes the "science" of the human sciences as the absolute paradigm of truth. This position entraps Dilthey's philosophy within the parameters of positivistic science as the sole source and sustainer of truth. Thus for Dilthey, as well as for Husserl, the science of the human sciences seems more primordial than philosophy itself. For Heidegger only Dilthey's close friend and associate, Count Paul Yorck of Wartenburg, could fill in the gaps in both Husserl's and Dilthey's joint unexamined dependence on scientism's apodicticity. Specifically, Yorck criticized Dilthey for failing to emphasize ". . . the generic difference between the ontic and the historical" (*SZ*, p. 404). Heidegger seized upon this difference as a radical ontological difference—on providing full access to the question of Being. Yorck's insight thus led Heidegger to grasping the essential distinction between the ontic realm of beings (*Seiendes*) and the historical realm of human being (historicity). But this distinction required a more original unity into which both the ontic and the historical could be unified and grounded. The impact of this insight on Heidegger is neatly summarized by J. L. Mehta:

> For the historical (as distinguished from the ontic) to be conceptualized philosophically, it is necessary that both the historical and the ontic should be comprehended under a more original unity making it possible to compare and contrast them with each other. This can be done only when it is realized that the question of historicity is an *ontological* question about the way historical existence is constituted, that the problem of the ontic is likewise an ontological problem about the constitution of non-human essents, of *vor-handen* entities in the wide sense and that the ontic comprises only one domain of what is./26/

Having thus acquired the necessary tools for distinguishing the ontic from the historical modes of being, Heidegger still required a direct means of approaching the question of Being. This requirement could be met only by finding a key and paradigmatic being, one that was at once both ontic and ontological. For Heidegger there was only one such entity: man himself. But to see man as the only ontico-ontological being required a precise phenomenological description of man in terms of his precise ontico-ontological characteristics. Thus man, as the only being which has access to Being, is characterized as a singularly unique type of Being called "Dasein." Hence, to understand how Dasein can be at once ontic and ontological, to understand how Dasein can even ask the question of Being to begin with, we must first understand the ontic everyday characteristics of Dasein. Here Heidegger was influenced in a fourth way by Søren Kierkegaard, whose approach to primordial Angst was discussed in our previous chapter.

As will become apparent, Kierkegaard's influence on Heidegger was delimited to illuminating the ontic properties of of Dasein (if Heidegger is to be taken at his word in *Being and Time*)./27/ Heidegger borrowed several key concepts from Kierkegaard to fill in what he called the "*Existenziell*" or ontic dimension of Dasein's mode of being. And, as we would expect, he acknowledged his debt to Kierkegaard. Among these were the familiar themes of situation, resoluteness, choice, death, authenticity, repetition, possibility, the anonymous "they," and, of course, the phenomenon of Angst itself./28/

To be sure, the relationship between Heidegger and Kierkegaard is extremely subtle and worthy of a full-dress study in its own right. This task has been undertaken with some degree of success by Michael Wyschogrod in his book *Kierkegaard and Heidegger*./29/ Our purpose here is not to comment further on their relationship; rather we may simply observe that Kierkegaard was indeed a major influence on Heidegger's ontic interpretation primordial Angst, but Heidegger's ontological concerns took him far beyond Kierkegaard's theological and psychological treatment of Angst. We shall, however, have occasion to examine this relationship in greater detail in Chapter V, where such comparisons become useful.

Looming large behind each of these specific influences on Heidegger's early mature period/30/ was the presence of two other figures who dominate *Being and Time*. These are Immanuel Kant and Franz Brentano. As a student at Freiburg Heidegger clearly was indoctrinated with the spirit of Neo-Kantian philosophy by Heinrich Richert, Heidegger's philosophy professor, who was himself a Neo-Kantian. This meant that Heidegger's first attempts to penetrate philosophy took place within a Neo-Kantian interpretation of the problems of values on the one hand, and of epistemology on the other. Heidegger's central ontological problem, therefore, was simply assumed to have been dissolved permanently by Kant himself./31/ Against this traditional mainstream Heidegger's Kant interpretation argued persuasively that Kant's purpose in writing the *Critique of Pure Reason* was not primarily epistemological; rather it was ontological. This Kant interpretation, later published as *Kant and the Problem of Metaphysics*, was conceived during 1925–26, several years prior to the publication of *Being and Time*, and was meant to be a first salvo against the tradition of Western metaphysics, but never saw print as a part of *Being and Time*./32/

In his Kant interpretation Heidegger struck deeply at Neo-Kantianisms. Specifically, he argued that Kant's purpose in the first *Critique* was to lay the groundwork for metaphysics; that is, not the *metaphysica specialis* of the schools (i.e., theology, cosmology, and psychology), but rather *metaphysica generalis*, the question of ontology. This *Being and Time* may be interpreted as a return to this fundamental Kantian problem, carried out as an inquiry into the general a priori conditions of the possibility of understanding and interpreting Being.

None of this, however, would have been possible without Franz Brentano's seminal influence on the young Heidegger. As we saw above, Brentano's dissertation gave rise to the abiding theme of Heidegger's thought, the question of Being. But Brentano influenced Heidegger in other, more subtle ways. First, from Brentano Heidegger learned to understand and appreciate the nuances of Greek philosophy. This led ultimately to the discovery of Truth as *aletheia*, another cornerstone of Heidegger's thought. Then, secondly, it was from Brentano that Heidegger learned an appreciation for neo-Scholastic philosophy in general and of St. Thomas and Duns Scotus in particular. Brentano's doctrine of intentionality is unquestionably related to the neo-Scholastic interpretation of intentionality, a point which neither Husserl nor Heidegger could possibly overlook.

Thus Brentano stands behind much of Heidegger's early mature thought, secretly moving it toward a confrontation with Neo-Kantian doctrines, urging it toward the Truth of Being as *aletheia*, and securing it in a neo-Scholastic tradition of which Brentano himself was a part. But none of this speaks to the enormous influence Brentano had over Husserl, the founder of transcendental phenomenology. And so in an indirect way as well, Brentano's influence went through Husserl to Heidegger, urging Heidegger to inquire into the question of Being in a new and primordial way not open to the Neo-Kantian tradition that dominated German philosophy prior to *Being and Time*.

This brief sketch by no means exhausts the many important influences on Heidegger during his early mature period, a complete listing of which would have to include Nietzsche and Pascal, to whom we shall return in Chapter V. But the six thinkers discussed here made far-reaching, profoundly deep, and therefore *major* contributions to Heidegger's thought during this period; contributions without which *Being and Time* could not have been conceived, much less written. Still, we concur with Mehta's observation that *Being and Time* is decidedly *not* simply a continuation of the agenda of these six thinkers; nor is it specifically a deepening of their individual perspectives. Rather, their joint influence, the *confluence* of their impact on Heidegger, allowed him to hermeneutically deepen the one question which concerned him: the question of the meaning of Being./33/

The Hermenuetical Task of Being and Time

We cannot overemphasize the fact that *Being and Time* is at root a foundational inquiry into the question of the meaning of Being. The analysis of Dasein that takes up the bulk of the published portion of *Being and Time*, therefore, is clearly a preliminary dimension of this essential question. So while *Being and Time* may *seem* to concern itself with philosophical anthropology, couched in the raiments of what Heidegger called "fundamental ontology," it is first, last, and always directed to the ontological problems raised by the question of Being. Hence it is clearly not a work of "existential thinking" as that term has been used by the followers of Jean Paul Sartre; neither is it a work of *Existenzphilosophie* as *that* term is used by the followers of Karl Jaspers. In fact, in a letter to Professor Arthur H. Schrynemaker, written in October of 1966, Heidegger tersely states: "Today it is hardly necessary anymore to remark explicitly that my thinking is a matter neither of Existentialism or Existenzphilosophy."/34/ Of course, this is but an expanded re-echo of earlier repudiations of "Existentialism" given both in the *Letter of Humanism* and Heidegger's work on Nietzsche./35/ But why are all of these disavowals necessary to begin with? The answer has to do with the manner in which Heidegger carries out his "fundamental ontology" or analysis of Dasein's ontological structures. This analysis takes its launching point form the ground definition of Dasein as that being indeed the *only* being, whose existence precedes its essence. This formula, "*Dasein's* existence precedes its essence,"/36/ has become as much the motto of "Existentialism" in Sartre's sense as "to the things themselves!" became for Husserl and phenomenology. So perhaps what is required here is to unpack the meaning of Heidegger's characterization of Dasein as the only being that is at once ontical and ontological, and whose existence as a *mode of being* precedes its essence as a being among other beings in the world.

What then is this curious being called "Dasein?" Heidegger himself answers:

> If the question of Being is to be precisely formulated and rendered completely transparent, then any elaboration of it . . . requires an explication of how Being is to be viewed, how its meaning is to be understood, and conceptually grasped; it requires preparing for the possibility of choosing the proper being as our paradigm, and working-out a genuine access to this being. [But] looking at, understanding and conceiving of, choosing, and making accessible are each constitutive types of questioning, and are thus themselves modes of Being of a particular being; which, as questioners, we ourselves are. Accordingly, to work out the question of Being means to render transparent a questioning being in its being. . . . *This being which we ourselves are, and which has questioning as one of its possibilities of Being, we shall designate by the term Dasein.* (SZ, p. 7)

Thus Dasein is the only being that can question Being, that has an abiding interest in Being, and that is capable of *dis*-covering hermeneutically the hidden meaning of Being. All of this is possible only because Dasein, as a questioner, has a privileged access to Being through its ability to ask questions.

But how then are we to grasp Dasein, this being which we ourselves are, or more phenomenologically precise, "I myself am,"/37/ and, moreover, where does the phenomenon of primordial Angst fit into such a grasp?

Heidegger addresses the question by suggesting that Dasein consists of four existential or ontological structures: (1) understanding (*Verstehen*), (2) predisposition or attunement (*Befindlichkeit*), (3) speech or discourse (*Rede*), and (4) what Heidegger calls "Fallenness or Forfeiture" (*Verfallen*). This fourth ontological structure makes possible a privative mode of Being-in-the-world, namely, Dasein's inauthentic mode of Being. Specifically, the absorption of Dasein into the "theyness" of the everyday world/38/ is revealed *ontically* through the phenomena of "chatter," or empty talk, "curiosity" or empty inquisitiveness, and "ambiguity" or empty understanding. Each of these are present in Dasein's everyday mode of Being.

Heidegger repeatedly insists that Fallenness or Forefeiture is not anything like a fall from a "purer" or "higher state of Being." Neither is it a display of something like man's inherent evil (whatever that might be). Rather, Fallenness or Forfeiture is co-equal with understanding pre-disposition and speech as the ontological structures or a priori conditions of Dasein's Being. Heidegger concludes from this analysis that fallenness is the ontological ground of Dasein's seeking out inauthentic modes of Being-in-the-world as the source of its everyday comfort and security. He gives an unintentional and certainly ironic example from his personal experience, in describing how it was that *Being and Time* came to be published.

> "Professor Heidegger—you have got to publish something now. Do you have a manuscript?" With these words the dean of the philosophical faculty in Marburg came into my study one day in the winter semester of 1925–26. "Certainly," I answered. Then the dean said: "But it must be printed quickly." The faculty proposed me *unico loco* as Nicolai Hartmann's successor for the chief philosophical chair. Meanwhile, the ministry in Berlin had rejected the proposal with the explanation that I had not published anything in the last ten years.
>
> Now I had to submit my closely protected work to the public. On account of Husserl's intervention, the publishing house Max Niemeyer was ready to print immediately the first fifteen proof sheets of the work

> that was to appear in Husserl's *Jahrbuch*. Two copies of the finished page proofs were sent to the ministry by the faculty right away. But after some time, they were returned to the faculty with the remark: "inadequate." In February of the following year (1927), the complete text of *Being and Time* was published in the eighth volume of the *Jahrbuch* and as a separate publication. After that the ministry reversed its negative judgment half a year later and made the offer of the chair./39/

Heidegger's disclosure of his "closely protected work," a work not only incomplete but incompletable in principle,/40/ displays how by seeking Hartmann's chair, he himself gave into the "theyness" as a condition of his eminent promotion. Needless to say, the results of his premature disclosure of *Sein und Zeit* for the sake of a desirable appointment displays Heidegger's own "inauthentic" side.

It is Fallenness, then, that gives inauthentic Being-in-the-world a special quality, ". . . a guarantee of the trustworthiness, genuineness, and fullness of all possibilities of being."/41/ However, in the phenomena associated with Fallenness (that is, the ambiguity, curiosity, and chatter) we notice a kind of deep and ontically inexplicable uneasiness: a feeling of agitatedness or alienation from our true selves./42/ This uneasy, agitated alienation, rather than spurring Dasein [as myself] to explore the depths of such a feeling, forces me to turn away from agitatedness, to shun and cover over the alienation, so as to seek comfort and security in my everyday Being-in-the-world. Ontically, "procrastination" is an example of such a drive to overcome the alienation, "refusing-to-consider" is yet another, "temper tantrums" are yet a third. Heidegger explains that this "comfort-seeking" has to do with the abyss metaphor we introduced in our first chapter.

> The phenomena of temptation, comfort, alienation and self-entanglement . . . characterize the special kind of being that belongs to fallenness. We call this agitatedness of Dasein . . . its "headlong plunge." Dasein plunges *from* itself into itself, into the abyss, and emptiness of inauthentic everydayness. But this plunge remains hidden from Dasein due to its public interpretation; so much so in fact, that it is interpreted as advancement or luring concretely. (*SZ*, p. 178)

Here the familiar echo of earlier chapters re-echoes: As Dasein I plunge into an abyss of ontological meaninglessness. Like

Kierkegaard's "dizziness at the face of the abyss," I experience a whirling and continuous dizziness at the edge of the abyss. Nonetheless, this whirling dizziness is a major connection point between the authentic and inauthentic modes of my Being-in-the-world. Therefore we must ask: what is the a priori condition that makes the dizziness and whirl possible? Clearly, in my ontic everyday mode of Being-in-the-world, fallenness is not revealed to me in any overt way. Thus my access to Being that makes dizziness possible must be grounded in one of my three other ontological structures. It seems clear that of these three the *only* candidate that can make dizziness possible is pre-disposition or attunement (*Befindlichkeit*). The pre-disposition I have toward Being that undergirds my everyday curiosity, chatter, and ambiguity reveals that such everyday phenomena are but inauthentic modes of Being-in-the-world. Moreover, pre-disposition or attunement reveals this mode with a vengeance that strikes me dumb.

We thus come to the connection point between fallenness and pre-disposition or attunement. Since we have explored fallenness in some detail, we must now return to a broader discussion of pre-disposition. For with examining pre-disposition we come at last to the immediate horizon of primordial Angst in *Being and Time*.

The Last Horizon—the Ontological Structure: "Pre-Disposition"

In the general introduction we observed that there are at least three equiprimordial structures constituting Dasein's mode of Being. These were: (1) understanding, (2) pre-disposition (or attunement), and (3) speech (or discourse). We also discussed in some detail how each structure displays a manner of Dasein—which is its disclosedness.

We now have at our disposal both the opportunity and the tools to examine the one ontological structure that concerns us primarily, namely, "pre-disposition," or "attunement." Angst, as the reader will no doubt recall, is the primordial pre-disposition or attunement (*Grundbefindlichkeit*) of Dasein. So to clearly understand Angst we must first deal with Heidegger's discussion of pre-disposition in general.

In the first place we may observe that at its Germanic root *Befindlichkeit* means "to find oneself in a situation," "to be attuned to the world" in a pre-reflective, non-thematic way which "pre-disposes" me to the ontic world of my circumspect preoccupations. In the everyday world I experience my pre-disposition through ontic moods (*Stimmungen*). And, of course, it is Dasein's ontological pre-disposition or attunement to the world that makes moods possible. As Dasein I always find myself in a specific mood: a good mood, a bad mood, a curious mood, a frivolous mood, a contemplative or reflective mood, and so on. But no matter what mood I am pre-disposed to at the moment, I am always *in* a mood of one type or another. This shows that Dasein is first and foremost a being that is delivered over to Being without much choice in the pre-disposition of its deliverance. I find myself in a factical/43/ situation I did not create. I am, I exist factically (as a naked and brute fact). This fundamental recognition comes over me. It dawns on me through my moods, revealing how I am attuned or pre-disposed to the world on a day-to-day basis, and that I *am* so attuned.

This "being-attuned-to-pre-disposition"/44/ is therefore my primary access to the world, which when taken in tandem with understanding and speech constitute the ontological foundations of my Being—the very foundation which undergirds my ontic, work-a-day preoccupation of ontic life.

To display the far-reaching meaning of pre-disposition in our ontic or everyday mode of being, Heidegger explores how pre-disposition undergirds the ontic phenomenon of fear./45/ Heidegger later uses this *phenomenal* analysis of fear as the precedent for his phenomenological analysis of primordial Angst. Thus his fear analysis has two purposes: (1) it discloses fear as an ontic or inauthentic mode of Being-in-the world, and (2) it sets the stage for how Angst, as the a priori condition of fear, is to be analyzed.

As Kierkegaard before him, Heidegger observes that in order for me to fear, there must be a specific threat to me, one coming towards me. While this feature of fear was about as far as Kierkegaard went, it is not far enough for Heidegger's quest to penetrate the very essence of fear. Heidegger separates the phenomenon of fear into three components. We must grasp each component, each must be clearly understood, if we are to grasp fully fear's phenomenal or ontic meaning. These three parts are, in brief: (1) the "before what" (*das Wovor*) element, or the nature of the threat to me; (2) the fearsome experience itself (*das Furchten*); and (3) the "about what" (*das Worum*)/46/ or the who-gets-threatened element. Since these same elements will be used again in Heidegger's ontological analysis of Angst, we might find it helpful to see how he employs this structure to analyze ontic fear.

First, the "before what" element refers to the "being" threatening me. We must, says Heidegger, discuss at least six subelements to tune finely the "before-what" of fear: (1) What threatens me has the *character of harmfulness* (*Abtraglichkeit*). (2) What threatens me has a distinct range of harm that can be done to me. Thus its harmfulness is revealed as coming from a definite region in my world. (3) The region from which the harmful comes is well known to me and now has something eerie (*geheuer*) about it. (4) The harmful is not yet within range of attack, but it is drawing closer, radiating its harmfulness; this accounts for its threatening manner. (5) The harmful is not so far off that it can be neglected; rather it is imminent and dangerously close. (6) Finally, the fact that the harmful has not yet attacked means that it *might* not attack. Yet this exact possibility serves only to heighten the intensity of the fear./47/

Secondly, "the fearing itself" (*das Furchten*) element occurs when I allow what is threatening to matter to me. I interpret something as "fearsome" because being fearful is an internal possibility of my ontological pre-disposition to Being. It is precisely because of this pre-disposition or attunement to the world that I can recognize something as a threat./48/

Third, the "about what" or the "what-is-it-that-gets-threatened" element is Dasein's own potentiality for *well* being. Only as Dasein can I be afraid, since I am the only being for whom Being is an issue. Thus fear discloses Dasein in a privative way, making conspicuous the need for everyday comfort and security by the absence of such comfort and security.

Heidegger takes great pains to show how "fearing for [the sake of] others" (*SZ*, p. 141) is radically grounded in being-afraid for oneself. He also show that the fear-associated phenomena such as alarm, dread,/49/ and terror are each simply variations of basic ontic fear.

> There can be variations within the constitutive moments of the full phenomenon of fear . . . resulting in differing ways of being fearful . . . [For example,] in so far as that which threatens, (in its mode of not right now

> but at any second) suddenly breaks into our preoccupied being-in-the-world, fear becomes *Terror* (*Erschrecken*). . . . The before-what of terror is for the most part well known and familiar. But if, on the other hand, that which is threatening us is something completely unfamiliar to us, then fear becomes *Dread* (*Grauen*). Moreover, where the threatening has the character of being dreadful, and at the same time it is encountered suddenly, then fear becomes *Horror* (*Entsetzen*). (*SZ*, p. 142)

We see from this passage, that dread, horror, and terror are *modes* of fear rather than of primordial Angst itself. Since primordial Angst is the ontological a priori condition of the possibility of fear, Angst must likewise be the a prior condition of dread, terror and horror as derived modes of fear. Thus, primordial Angst can never be directly equated with dread, terror or any other fear-related ontic phenomena, since such an equation violates the radical differences between the ontological and ontic realms. Not only that, but the related phenomena of timidity (*Schüchternheit*), awe (*Schue*), anxiety/50/ (*Bandigkeit*), and astonishment (*Stutzigwerden*) are *all* ontic phenomena grounded in the phenomenal paradigm of fearfulness (*Das Furchten*). Of course, the explicit reference to "anxiety" here should not be overlooked.

With this analysis as a guide, we may now proceed to primordial Angst itself. For in laying bare the ontic or everyday structures of pre-disposition as a constitutive element of Dasein, we have come at last to the final horizon of primordial Angst—the ontological dimension of pre-disposition. We may now see the role primordial Angst plays in Heidegger's thought: Angst is the most original, the furthest-reaching, and indeed the only means whereby Dasein has access to understanding the ontological significance of Being-in-the-world.

Angst in Being and Time

Heidegger, after having displayed the ontic significance of Being-in-the-world, now seeks for a means to hermeneutically understand and interpret the ontological wholeness of Being-in-the-world. The key to such an understanding, he insists, is a *phenomenological* rather than a phenomenal interpretation of Angst. In section 40 of *Being and Time*, therefore, Heidegger reveals what we have called primordial Angst by contrasting Angst with the phenomenon of ontic fear discussed earlier. Primordial Angst as an ontological phenomenon is contrasted with ontic fear by using the same analytic structures used previously; (1) the "before what" (*das Wovor*) element, or the nature of what makes me as Dasein Angst-filled; (2) the Angst-filled experience itself; and (3) the "about what" (*das Worum*) element, or who is it that experiences primordial Angst.

Heidegger begins by discussing the relationship of forfeiture or fallenness to both fear and primordial Angst. Heidegger reasons that fallenness implies a "fleeing from." While in fear I flee from what threatens me, in primordial Angst no specific being threatens me. Thus, fallenness shows me turning away from *myself* in primordial Angst; that is, turning away from my authentic mode of Being. What does turning away and fleeing reveal phenomenologically to Dasein? A closer examination of the structures of primordial Angst may be useful here in answering this question.

First, the "before-what" element of primordial Angst is manifestly not any specific being, nor is it another Dasein along with me in the world. Were that the case, I would experience fear rather than primordial Angst. Let us, then, retrace the subelements introduced in the analysis of fear as they are now applied to primordial Angst: (1) What threatens me as a Dasein in fear, namely, the harmful being with which I am concerned, is completely absent in primordial Angst; for in Angst I am Angst-filled precisely because *I cannot determine what it is that threatens me.* (2) Moreover, unlike fear, in which I detect a threat as coming from a well-known region, in primordial Angst I experience the threat as coming from nowhere specifically (i.e., no specific region). (3) In fear, that well known *region* takes on an "eerie" quality to it as it threatens; but in primordial Angst the world itself becomes eerie; nothing remains familiar. (4) Whereas in fear I feel that what threatens is not yet here but is drawing nearer, in primordial Angst the amorphous threat is already *here*; it is non-directional and non-temporal in this sense. (5) In fear I detect the threatening as being so close that it demands my immediate attention; but in primordial Angst the threat is neither close nor far, neither here nor there: it surrounds me completely. (6) Finally, during fear I may hope that what threatens will not attack, since it has not attacked as yet.

Not so in primordial Angst. In Angst I am attacked by nothing, and yet I feel myself under constant attack from no-where and from no-thing. Thus *it is the world itself that is the "before what" of Angst*, experienced ontologically as a silent apprehension of my individual being as a thrown Being-in-the-world. Says Heidegger:

> What oppresses us is not this or that much less the sum of the things together; rather it is the possibility of the referential context of world in general; that is, the world itself. As soon as Angst has abated we are accustomed to saying in everyday language, "it was really nothing." *What* it was, was ontically hit upon in this manner of speaking. . . . Thus, if the no-thing; that is, the world as such, shows itself as the "before-what" of Angst, it may be said then that the "before-what" that Angst is *Angst-filled* about, is Being-in-the-world as such. (*SZ*, p. 187)

Thus the apprehension of primordial Angst itself discloses the world through Dasein's pre-reflective experience of its world-hood: namely, the outermost phenomenological horizon, the no-thingness that lights up in the totality of our circumspect concerns (*Bewandtnisganzheit*). What Dasein apprehends in primordial Angst is the vacuousness of the world as beings disappear into the abyss of meaninglessness. "The world," says Heidegger, "has nothing more to offer, much less so the being-with of others" (*SZ*, p. 187). In robbing me of comfortable self-understanding, primordial Angst throws me back upon my potentiality for Being-in-the-world. Primordial Angst therefore isolates (*vereinzelt*) me, it frees me to project myself upon my authentic possibilities, making it possible for me to choose myself concretely. This is not simply a psychological and ontic choice à la Kierkegaard. Instead, for Heidegger, primordial Angst displays in a pre-reflective way the *ontological* conditions undergirding *any choice whatsoever*. In contrast to Kierkegaard (for whom Angst was the *ontic* possibility of possibility), Heidegger sees primordial Angst as the *ontological* possibility of possibility: the "ground" of ontological freedom./51/

Third, the "about what" or the "who-it-is" that is stricken in primordial Angst is, of course, Dasein. Heidegger spares no small effort to show that the "before-what" element (*das wovor der Angst*), and the "about-what" element (*das worum der Angst*) are *identical*. This selfsameness, Heidegger argues, assures us that primordial Angst is a *distinctive* disposition, one capable of a full-dress phenomenological analysis and hermeneutical interpretation.

The ontic or experiential manifestation of primordial Angst, according to Heidegger, is "uncanniness" or "eerieness" (*unheimlich*). Strictly speaking, we do not directly "experience" primordial Angst. Angst is the underlying condition which makes the "eerieness" experience possible. Hence, what I "experience" is a marked sense of "uncanniness" which overpowers me. So when such uncanniness overtakes me, I flee *into* the tranquil, public, and everyday world of ontic familiarity. Simultaneously, of course, *I flee from* the daunting uncanniness.

Ontic uncanniness can strike anytime, in any place. It pursues Dasein relentlessly in the darkness as well as in the full light of day, in the battlefield or in a loved one's arms. As that which constitutes Dasein's being, the primordial pre-disposition of Angst is itself a concrete mode of Dasein's Being-there (*Dasein*)./52/ Like some ontological shadow of our Being, Dasein can never, under any circumstances, escape Angst. Hence, we see that the "feeling" of fear is possible only as a fallen-into-the-world mode of primordial Angst—an Angst which, when it becomes fear, is hidden from itself.

While authentic or primordial Angst is rare, even "physiologically conditioned anxiety" is at root an ontologically grounded problem. For only because I, as Dasein, am Angst-filled in the ground of my Being *can* ontic or psychological anxiety manifest itself physiologically. The most distinctive characteristic of primordial Angst is its power to isolate men from the "theyness" of the world. Angst makes me *solus ipse*. Primordial Angst, therefore, strips me of my illusions; revealing the world in an undisguised manner; revealing that I cling inauthentically for the most part to the circumspect world for security, a world of ontic familiarity, a world in which I am comfortable.

Throughout the remainder of *Being and Time*, Heidegger continues to build upon his basic Angst interpretation, showing how primordial Angst reveals *care* as the basic meaning of Dasein's being./53/ In section 41, for example, Heidegger argues that since primordial Angst reveals my potential to become my future possibilities, it follows that an essential part of my being must consist in an anticipatory or a future orientation. I am, therefore, always anticipating my future. I am already thrown into a world I did not create. So I have a "past" orientation toward my Being as well as a future orientation. Primordial Angst discloses the full ontological structure of Being-in-the-world, phenomenologically a unity of three temporal characteristics: (1) anticipating myself as a future orientation of Being, (2) being already thrown into the world as the past orientation, and (3) dwelling in the world along with other beings as the present orientation. Heidegger observes in summary: "This Being fills in the signification of the term *care* (Sorge), which is employed in a purely ontological manner" (*SZ*, p. 192). Care, then, is the a priori condition of the possibility of Dasein's Being-in-the-world, from the overreaching perspective of *Being and Time*.

In section 50, "Sketch of the Ontological Structure of Death,"/54/ Heidegger reveals primordial Angst as Dasein's primordial pre-disposition in the context of apprehending my ultimate ontological potentiality—death. First, says Heidegger, the ontological structure of death is Dasein's ". . . ownmost, nonrelational, and non-surpassable possibility; one which is furthermore a distinctively impending event."/55/ Second, as Dasein, I do not generally have any explicit or even a theoretical knowledge of my impending Death. Rather, through primordial Angst the ontological significance of death is revealed to me. Says Heidegger:

> Thrownness towards death shows itself originally and forcefully in the disposition of Angst. Angst in the face of death is Angst (standing) before one's ownmost, nonrelational, and non-surpassable potentiality-for-Being. The "before-what" of this Angst is Being-in-the-world itself; its "about-what" is simply *Dasein's* potentiality-for-Being. We must not confuse Angst in the face of death with a fear of becoming decrepit. Angst is not some arbitrary and random feeling of faintheartedness in someone; rather it is a revelation through *Dasein's* most basic disposition that *Dasein* is a thrown being existing *toward* its own end. (*SZ*, p. 251)

Thus, primordial Angst reveals not only my basic freedom to become my potentialities, as we saw in Section 40, but Angst likewise reveals the ultimate limit to those potentialities. These limits I must face alone. Death is my ultimate destination. Viewed from the perspective of the everyday world, to show "anxiety" in the face of death at the hands of a firing squad, for example, is considered cowardly (*SZ*, p. 254). This is because the anonymous "they" interpret such "anxiety" as fear in the face of a specific impending event, the unknown that follows the rifle's discharge. Such misunderstood Angst is interpreted as the "faint-heartedness" mentioned in the passage above./56/ Needless to say, this is completely beside the point in discussing the foundational dimension of primordial Angst.

Perhaps the most significant aspect of the relationship between death and primordial Angst is the following: Being-toward-death is essentially an authentic "living-through" of Dasein's primordial pre-disposition, Angst. Specifically, since every understanding (*Verstehen*) must be accompanied by an equi-primordial attunement or pre-disposition, (*Befindlichkeit*) the primordial pre-disposition, Angst, allows me to ontologically interpret death as the final horizon of all possible potentialities of my own Being. At the same time, however, primordial Angst reveals death, as my ownmost, non-relational, non-surpassable potentiality, to be radically ambiguous with regard to *when* it will occur. So then, death always remains ahead of me as a constant but non-specific threat, obscurely looming just over the horizon of the future as nothingness for Dasein. Heidegger states:

> . . . Angst is the pre-disposition capable of holding open to Dasein the constant and sheer menace arising from Dasein's isolated being. In Angst Dasein finds itself face to face with the Nothingness of the possible impossibility of its existence. . . . For this reason, the basic pre-disposition of Angst belongs to the very depths of Dasein's self-understanding. (*SZ*, p. 266)

Another facet of primordial Angst is the "call of conscience." The "call" summons Dasein back from the "theyness" of inauthentic existence. This "call of conscience" summons me to the resoluteness (*Entschlossenheit*) required for choosing my authentic self./57/ Primordial Angst plays a key role in both dimensions of Dasein's wholeness and authenticity. Specifically, through primordial Angst I must, if I am to be authentic, (1) hearken to the silent call of conscience (*das schweigende Reden des Gewissens*), (2) accept ontological guilt, and (3) will to become authentically resolute.

We must consider each of these briefly in relation to primordial Angst: (1) The call of conscience comes from the depths of my being in the everyday disposition of uncanniness. The call is attuned by Angst, therefore, and ". . . is what makes possible at the outset my ability to project myself on my ownmost possibilities" (*SZ*, p. 277). (2) My ontological guilt comes also from the depths of my being, as a thrown Being-in-the-world. Thus ontological guilt (which should not be confused with ontic guilt in a psychological or religious sense à la Kierkegaard) is the content of the call, revealed through primordial Angst. (3) Heidegger defines resoluteness as a ". . . wordless, Angst-ready, self-projection upon one's ownmost Being-guilty" (*SZ*, p. 297). Resoluteness allows me to become authentic by forcing me to face primordial Angst. In facing primordial Angst I (a) hearken to the call of conscience, (b) accept my disclosure of guilt, and (3) accept fully the resoluteness to become a genuine self.

Finally, this entire analysis, with its manifold descriptions of Angst's meaning, is recast in the light of Dasein's temporality. To appreciate fully how temporality throws completely new light on primordial Angst, we must first see how Dasein is essentially a temporal being to begin with.

Dasein is a being whose essence *is* its existence./58/ Again, Dasein's existence is constituted by four ontological structures: understanding, attunement or pre-disposition, fallenness, and speech. Primordial Angst is, of course, the most basic attunement of the second structure, "pre-disposition." But since these ontological structures constitute Dasein's being equally, they cannot be completely distinguished from one another. Hence, every understanding has its pre-disposition, every pre-disposition its fallenness, every fallenness its spoken articulation. What, then, is the temporal signification of this complex being called Dasein, and how is temporality related to primordial Angst? Temporality has its own ontological structure. Heidegger calls these "moments" or "temporal ecstacies." The three ecstacies are familiar; they are the future, the present, and the past. One moment or ecstacy cannot be fully distinguished from the others any more than one of Dasein's ontological structures can be isolated from the others.

To grasp the primary temporal orientation of Dasein's primordial pre-dispositions is to grasp that both fear (as an inauthentic pre-disposition) and primordial Angst (an authentic one) are chiefly grounded in the "has been" (*ist gewesen*) of the past.

In the first place, says Heidegger, having a mood as a kind of pre-disposition, brings me, as Dasein, face to face with my thrownness. Hence, ontologically speaking, moods are a result of my "having-being-thrown" into the world. The "having-been" structure of thrownness reveals the primary temporal ecstacy of my pre-dispositions to be the past. I can be brought face to face with thrownness only if in the essence of my being I am my "has-been."

First, fear is an inauthentic pre-disposition. How then does its ontological significance lie in what "has been"? Fear is possible only on the basis of a specific threat. So to fear I must let something threaten me from within my world. And I must await it as it comes closer to me. Thus fearing is "the expectation of an approaching evil (*malum futurum*)" (*SZ*, p. 241), a kind of reaching-out into the future which makes me forget my own being. In fear I focus upon what it is that threatens. But reaching-out into the future is itself possible only because I am grounded in my past. I forget that thrownness is the original condition of my Being-in-the-world. In fear I forget my self so as to escape through flight. What is disclosed in the *having-been* forgotten of thrownness is what I call "the past."

Second, primordial Angst is an authentic pre-disposition. Indeed, it is Dasein's most authentic pre-disposition. As such Angst brings me as Dasein into confrontation with thrownness, disclosing in that confrontation the uncanniness of everyday Being-in-the-world. Heidegger shows that the "before-what" and the "about what" structures of primordial Angst are one and the same as Dasein's potentiality for Being in-the-world. Clearly, then, in primordial Angst there is no specific threat to threaten me. Rather, the threat comes from within me, as I am overwhelmed by an apprehension that the environing world is losing its familiar significance.

A *Nothingness* is announced through the phenomenon of primordial Angst. The world and everything in it loses its contextual meaningfulness and the abyss of meaninglessness, an ontological vacuousness overtakes all beings in the world, leaving me likewise transfixed. Hence, I am repeatedly thrown back on my isolated and naked existence as a thrown Being-in-the-world. I repeatedly recognize my thrownness as primordial Angst overtakes me. Hence I cannot forget my thrownness as I could in the ontic phenomenon of fear. On the contrary, I stand now *poised* for resoluteness in primordial Angst, remembering my thrownness with crystal clear-headedness. Primordial Angst, therefore, must be grounded in the temporal ecstasy of "having been" for any of this to be possible. Heidegger discloses the ontico-ontological significance of this fact: "the resolute person knows no fear, understanding precisely the possibility of Angst as a mood which does not hinder or confuse but rather liberates him *from* vain possibilities and permits him to be free for authentic ones" (*SZ*, p. 344).

With that observation Heidegger completes the final touches to his portrait of primordial Angst in *Being and Time*. True to his own phenomenological method, Heidegger has moved us several full sweeps around the hermeneutic circle, steadily deepening our awareness of primordial Angst's ontological significance and fashioning progressively vaster horizons for Angst's ontological interpretation.

Heidegger's exhaustive display in *Being and Time*, however, was not his final word on Angst. In fact, as we shall see in the next section, primordial Angst assumes an even broader horizon in Heidegger's thought after 1927, when he came to see that fundamental ontology, precisely because it uses the language of metaphysics, can never approach Being in itself./59/

As we will soon see, the new portrait of primordial Angst displayed by Heidegger in his 1929 Inaugural Lecture "What is Metaphysics?" has equally far-reaching consequences for philosophy as well as for the Human Sciences.

Angst in "What is Metaphysics?"/60/

We come now to Heidegger's final discussion of primordial Angst. In many ways, his Angst discussion in "What is Metaphysics?" reveals Angst's central role in Western thought much more clearly than Heidegger's *Being and Time* Angst interpretation. For in this Inaugural Lecture, primordial Angst is shown to be both the a priori condition of the possibility of finite human transcendence, and the condition of *any scientific knowledge whatsoever*. To be sure, this is a claim which if true is of monumental importance to human thought in general. Let us therefore examine this claim in greater detail.

Heidegger begins his lecture by observing that the question "What is Metaphysics?" has two essential components: (a) what the question concerns, in this case "Metaphysics," and (b) from whose standpoint the question is being asked, in this case "Dasein's." So, the question "*What* is Metaphysics?" must include Dasein as the questioner in what is being questioned. The reader will recall, from the section entitled "The Hermeneutical Task of *Being and Time*" above, Dasein was defined by Heidegger as that being which questions. Thus, in this address to the faculties and students of Freiburg University, Heidegger reminds these scholars that precisely *as* a community of scholars, the university's own Being is governed by the questioning of science. Given the basic truth of this statement, Heidegger asks: "What essentially happens to us, in the ground of our Dasein, when science becomes a passion to us?" (*WM*, p. 1/*G*, p. 103); for surely, science in its broadest sense has become the passion of those Dasein whose primary environing world (*Umwelt*) is the university.

To answer this haunting and fundamental question requires some in-depth analysis. Heidegger observes three elements to *Dasein's* scientific existence: (1) Scientific Dasein has a relationship to the world that allows him to seek out beings by making them objects of investigation. (2) In this activity, scientific Dasein submits to the things themselves so that they may reveal themselves through science. (3) In the pursuit of science, scientific Dasein, as the being that questions, actually "irrupts" into the whole realm of beings in a particular way, so that beings become illuminated—showing *what* they are as well as *how* they are. Over against this scientific relationship to the world, says Heidegger, there is really *nothing* to consider beyond Dasein's concern for beings and Dasein's irruption into the world of such beings. Hence, scientific Dasein must concern himself *only with examining beings, and beyond that nothing.*

Heidegger then asks the essential metaphysical question, the question that "goes beyond" (meta-) beings (physics): What about this "*nothing*" (*WM*, p. 3/*G*, p. 1205)? Clearly the nothing is rejected by science as a kind of cognitive vacuousness, which if taken seriously can only serve to undermine science's intellectual rigor. However, by steadfastly insisting that science's purview is restricted to beings alone, science *calls upon* the nothing; indeed *uses* the nothing as the cognitive background against which beings are revealed. Hence, concludes Heidegger, science tries to ignore the nothing but in the same breath it surreptitiously admits the nothing into its hallowed academic halls and laboratories. Says Heidegger: "With this reflection on our immediate existence as one determined by science, we find ourselves embroiled in the midst of a controversy. Through this controversy a question has already evolved which requires only an explicit formulation: How is it with the nothing?" (*WM*, p. 4/*G*, p. 106).

Heidegger then shows his audience that the nothing: (1) can not be an "object," because to make the nothing into an "object" expresses a contradiction, and (2) the nothing is beyond the purely logical negation of the totality of beings, that is, "non-being" pure and simple; for the conceded nothing is itself the a priori condition of the possibility of the "not" *and* of the logical activity of negation as well. Where then, asks Heidegger, are we to find the nothing? How, moreover, do we even know what to look for?

Perhaps we might start with examining ordinary experience: we *use* the term "nothing" everyday in our ordinary commerce with the world. We can even *define* the nothing as "the complete negation of the totality of beings" (*WM*, p. 6/*G*, p. 108). But this provides only a formal concept, not "the nothing" as science employs it. Nonetheless this definition reveals an essential clue, says Heidegger: in order to negate the totality of beings we must first encounter their wholeness in our daily lives; and, in fact, this is precisely what *does* occur when we find ourselves in the midst of this totality we call "the world." What we experience, then, is the "wholeness of Being." Says Heidegger:

> No matter how fragmented our everyday life appears to be, it still treats beings, no matter how obliquely, as unities of the whole. Even then—especially then, when we are not busy with things or with ourselves, we are overcome with this wholeness; for example, in genuine boredom. Boredom is still far off when we are merely bored with this book or that play, with that busyness or this indolence. Boredom breaks-in when "one is [profoundly] bored." Profound boredom, drifting back and forth through the abysses of human existence like a muted fog, draws all things, all men and oneself with it into a peculiar indifference. This boredom reveal beings as a whole. (*WM*, pp. 7–8/*G*, p. 110)

The wholeness of Being is likewise revealed as ". . . in the joy of the existential presence—and not simply the person—of a loved one" (*WM*, p. 8/*G*, p. 110). Only through such specific moods (*Stimmungen*) do we discover how we are in the world, as well as *that* we are along with other beings in the world. In moods the totality of the world is revealed to us, but the revelation simultaneously covers-over the nothing. There is only one key disposition, one special mode of attunement, that can reveal the nothing—the primordial pre-disposition of Angst.

In "What is Metaphysics?" Heidegger is very explicit about Angst's features. Primordial Angst, possesses the following characteristics: (1) It is not reducible to a common ontic "anxiousness" (*Angstlichkeit*) which must be grounded in "nervousness" (*Furchsamkeit*). (2) Angst is associated with a kind of calm, a quiet peace that pervades Dasein. (3) In Angst we "feel" uncanny. This uncanniness pervades all things, including one's Dasein. (4) Primordial Angst *reveals the nothing*. (5) In Angst we are suspended: we hover in suspense as the *totality* of the beings, their meaning and significance for us, slips away into the abyss. Pure being-there (*Dasein*) is all that remains: in Angst we are naked and vulnerable to the nothing. (6) Primordial Angst strikes us dumb, robbing us of our ability to say anything meaningful. We chatter compulsively, proving only the all-pervasiveness of the nothing peeking through our empty chatter. But in the vivid moments that follow this fleeting recognition of the nothing, we say that it was "really nothing" that troubled us. With these characteristics of primordial Angst, the discoverer of the nothing, Heidegger is prepared to deal in yet a clearer way with what can be said about the nothing.

In order to approach the nothing concretely, man must be changed, says Heidegger, into *Da-sein*, an event occuring each and every time we recognize our primordial Angst./61/ Heidegger implies that without primordial Angst man cannot become Dasein. In primordial Angst the nothing is encountered when the totality of beings slip away into meaninglessness. "In the clear light of the nothing of Angst," says Heidegger, "the original openness of beings as such arises: that they are beings—and not nothing" (*WM*, p. 11/*G*, p. 114). Because of primordial Angst, Dasein can grasp the things themselves. "Only on the ground of the original disclosure of the nothing can the Dasein of mankind approach and penetrate beings" (*WM*, pp. 12–13/*G*, pp. 114–15). This is the essence of science. Thus to be *Da-sein* means precisely "to be-held-out-into the nothing" (*WM*, p. 12/*G*, p. 115). Such a being-held-out-into (*Hineingehaltenheit*) means that *Da-sein* is already beyond the totality of beings, a condition we commonly call "transcendence." Without transcendence we could *never* relate to the totality of beings in a scientific or, for that matter, any other way. Neither could we relate to ourselves. Thus Heidegger concludes: "Without the original disclosedness of the nothing, no selfhood and no freedom [are possible]" (*WM*, p. 12/*G*, p. 115).

All of this gives rise to an important question: If transcendence means being held out into the nothing, and if, moreover, the nothing is disclosed only in primordial Angst, then does it follow that in order to exist at all we must constantly hover in primordial Angst? To this important question Heidegger answers yes, "Angst is there. Only it is sleeping" (*WM*, p. 14/*G*, p. 117). Why? Because Dasein falls into the world of public concerns in primordial Angst; indeed it is primordial Angst which impels Dasein into ontic preoccupation. In its nihilating essence, "the nothing directs us precisely toward being" (*WM*, p. 13/*G*, p. 116). So the nihilating activity of the nothing goes on ". . . continuously, without our everyday manner of knowing" (*WM*, p. 13/*G*, p. 116). The nothing shows daily through our glib ontic preoccupation from time to time, but for the most part remains hidden. In the activity of negating we see the more primordial nothing showing itself as the ontological condition behind negation.

> However frequently or in how many ways negation—explicit or not—permeates all thought, it is by no means the only fully compelling witness for the revelation of the nothing that essentially belongs to Dasein. For negation cannot claim to be either the only or even the leading mode of nihilating behavior in which Dasein remains shivering in the nihilation or the nothing. More abyssmal than the vapid aptness of rational negation is unbending opposition and the acrimony of abhorrence. A deeper answer is required to explain the sorrow of failure and the mercilessness of prohibition. More oppressive is the bitterness of privation. (*WM*, p. 14/*G*, p. 117).

Yet for all this, primordial Angst is not the opposite of gladness or simple joy, or even the tranquil ontic contentment of everydayness—provided that we are daring enough to face Angst openly. Primordial Angst is, says Heidegger, ". . . in secret alliance with the serenity and mellowness of creative yearning" (*WM*, p. 15/*G*, p. 118). But primordial Angst, even as it sleeps within Dasein's ontic mode of Being, can awaken at any moment, needing no stupendous event to awaken it. It is *always there* lurking just beyond the fringe of awareness, waiting to reach out to seize and transform us.

Hence the question "What is Metaphysics?" becomes answerable only with reference to Dasein's finite transcendence, a transcendence made possible only by primordial Angst. Primordial Angst reveals the nothing and makes metaphysics possible in the first place, a fact that the scientific community must recognize if it is to be true to its own intellectual roots. For although that community of scholars and scientists would prefer to ignore the nothing, it must recognize that *scientific* Dasein is itself possible only to the extent that it holds-itself-out-into the nothing in pure wonder of primordial Angst. Heidegger notes "the presumptive sobriety and the superiority of science becomes a laughable absurdity, if it does not take the nothing seriously" (*WM*, p. 18/*G*, p. 121). Furthermore, only because the nothing is revealed in primordial Angst can science make beings the object of scientific investigation; for without primordial Angst, no transcendence is possible. So without transcendence we, as scientific Dasein, could never "get beyond" the things themselves *in order to* understand them. Primordial Angst, therefore, is the root of science.

In the "Postscript"/62/ to *What is Metaphysics?* written in 1949, Heidegger clarifies some of his observations on primordial Angst, specifically in connection with the major objections to his Angst discussion raised by his detractors. These objections are: (1) The lecture makes the "nothing" the sole subject of metaphysics, leading to a philosophy of Nothingness—the "last word in nihilism." (2) The lecture raises an isolated and morbid mood, namely Angst, to the status of a key mood. But this mood, say the Heidegger detractors, is the "psychic condition of the nervous and of cowards . . ." (*Nachwort*, p. 101/*G*, p. 305). The resulting philosophy of Angst paralyses the will to act and undermines the stout-heartedness of the courageous. Finally, (3) the lecture declares itself against logic, arguing for a philosophy of pure feeling, which "endangers precise thinking and the certainty of action" (*Nachwort*, p. 101/*G*, p. 305).

Heidegger's response to objection (2) sheds considerable light on the other objections. Specifically, regarding the charge that the lecture raises primordial Angst to a key mood that paralyzes the will and undermines courageousness, Heidegger responds along the following lines of thought:

In the first place, no matter how hard science tries to find Being among the totality of beings, science can only come up with the sum of these beings, never Being itself. Only in the Nothing, does Being reveal itself as the essential "other" that is distinguished from beings. Moreover, *only* in primordial Angst can this revelation take place as the negation of beings, a negation, moreover, made possible by the Nothingness. Thus if we do not, out of sheer cowardliness, ". . . avoid [hearkening to] the silent voice that attunes us to the terrors of the abyss . . . we will hearken to the experience of Being, appropriated as the wholly 'other'" (*Nachwort*, p. 102/*G*, pp. 307–8), from which each being, indeed all beings, are distinguished.

Therefore, to detach primordial Angst from Being is to force Angst into an ontic, psychological *feeling* that is utterly beside the point of Heidegger's thought.

> Readiness for Angst is [the ability] to say "yes!" to the earnestness of things; to fulfill the highest demand which alone touches man's essence. Only man among all beings when called to be the voice of Being, experiences the wonder of wonders: *that* beings exist. Therefore the being that is called to the truth of Being is always attuned in a primordial way. The clear courage for primordial Angst vouches for the most mysterious of all possibilities: the experience of Being. For close to primordial Angst—as the terror of the abyss—abides awe. Such awe, illuminates and covers-over each dwelling place of mankind, within which he comfortably abides in the abiding. (*Nachwort*, p. 103/*G*, p. 307)

Angst interpreted as "anxiety" is simply a kind of psychological dread of dread. This ontic vision fails to grasp the essence of primordial Angst: a kind of ontological *courage*—the courage to face steadfastly and stoutheartedly the terrors of the abyss at the very heart of Dasein's ontological structure. Purely psychological anxiety, again as Heidegger showed earlier in *Being and Time*, is simply confused fear, a kind of inauthentic feeling that has nothing to do with primordial Angst. Heidegger concludes his discussion of Angst by noting that to the degree we degrade primordial Angst and its essential relationship to man and Being, we likewise degrade the essence of courage.

> In the abyss of terror, courage recognizes the virtually unexplored realm of Being: that openness into

> which each entity returns as what it is and what it can be. This lecture advocates neither a "philosophy of Angst," nor does it seek to give the surreptitious impression of a "heroic philosophy." Its only thought concerns that which dawned on the Western intellectual tradition at its very outset, but nevertheless remains forgotten—Being. (*Nachwort*, p. 103/*G*, p. 308)

With this observation we may conclude that Heidegger has worked his way through primordial Angst only to return to his point of departure: the Question of Being, a return that is consistent with his quiet passion for Hölderlin's point in his *Hymn to the Rhine*: ". . . for, how you begin is what you will always remain."/63/

Preliminary Conclusions

So by way of answering the above questions preliminarily we may conclude at this juncture several points requiring further analysis and discussion: (1) Heidegger clearly elevates primordial Angst from an ontic, psychological, phenomenal phenomenon to an ontological, a priori, phenomenological phenomenon. (2) In doing so Heidegger prepares the ground for two fundamental problems that go beyond his own analysis of primordial Angst—Angst in relation to a full-dress treatment of the Nothingness, and Angst in relation to ontological courage in the face of such Nothingness. (3) Heidegger brings the transcendental phenomenology of Husserl and the "Philosophy of Life" of Dilthey into concert with Kierkegaard's vision of Angst, displaying an ontically grounded approach to the ontological substructure of the phenomenon. (4) Heidegger accounts for understanding (*Verstehen*) as being more than merely cognitive by tying understanding to Angst as Dasein's basic or primordial pre-disposition. This accounts for our non-cognitive modes of understanding, a big step indeed. (5) Heidegger's use of primordial Angst seems to provide unity to the vast ontic diversity of experience, covering in the process contingencies such as fear, boredom, etc., which are not covered in previous attempts to seek a complete ontology. (6) Heidegger's concept of Angst in relation to Nothingness is what ultimately grounds the possibility of science.

Over against these accomplishments there stands an important set of concerns that require macro-hermeneutical discussion and examination if we are to judge finally Heidegger's contributions as going beyond simply historical significance. These concerns delve deeply into the ground of hermeneutical phenomenology itself.

1. Heidegger's account of the phenomenological phenomenon called primordial Angst may, after all, be an ontological construction which cannot *in principle* be grounded in ontic experience. There is a major leap, after all, between ontic "uncanniness" and ontological Angst. We can only assume Angst is there. We never experience it.

2. This gives rise to the much deeper problem of grounding Dasein's ontological structures, a problem which Heidegger himself admits takes precedence over ontological inquiry./64/ In *Being and Time* he asserts that the roots of the analysis of Dasein are purely ontic./65/ Heidegger himself had to abandon his own question of Being as a fundamental ontology because one cannot get to Being in general from Dasein's being.

3. Heidegger's analysis of Dasein's ontological structures are challenging and illuminating, even intellectually exciting. Certainly they are *creative*, and may be well worth our attention. But we should still ask with propriety: Do his descriptions and the underlying concept of Being to which they point adequately account for Dasein's general experience? And, moreover, is his account more economical than an alternative description? To this question we must add two subquestions: (a) Does Heidegger's own recognition that there could never be a second half to *Being and Time* suggest that he himself cut the heart out of his early mature thought? And, (b) does not Heidegger's elaborate revelation of Dasein's ontological structures add up to yet another instance of what Plato in the *Timaeus* called "a likely story"?/66/ Surely, the inconclusiveness of Heidegger's ontology, when taken in concert with what Caputo has revealed as Heidegger's mystical tendencies,/67/ lends credence to this interpretation. But, more importantly, if Heidegger asks us to think in a more primordial way, and if, moreover, this requirement is to have meaning and significance for us in general, there must be important ontical consequences for us as human beings, trying to find meaning in our own fragmented existence. Indeed, this may be the radical connection between hermeneutic phenomenology and pragmatism—a "cash value" to the question of Being. The major contribution of *Sein und Zeit*, then, may be ontic after all.

4. Heidegger's use of Gnosticism's language, symbolism, and

categories reveals a hidden dependence on that tradition which is never clearly brought to light in *Being and Time*. Furthermore, the gnostic-like notions of thrownness, the call, and of Dasein's inauthentic alienation from Being in everyday Being-in-the-world are taken as givens as Heidegger's analysis of Dasein and its ontological structures.

5. Finally, there is the question of the structural purpose of Heidegger's Angst. To deny that Angst plays an essential role in Heidegger's early mature period is wholly unsupportable. Certainly much more than simply just another "mood," Angst—primordial Angst—is Dasein's basic pre-disposition or mode of attunement toward the world. Moreover, Angst is at the very root of scientific Dasein's mode of Being. Thus Angst is for Heidegger the key means to disclosing: (1) how fear is possible, (2) how other moods are possible, (3) how Dasein's own Being in-the-world is possible, (4) how Being-toward-death is possible, (5) how anticipatory resoluteness is possible, (6) how finite transcendence is possible, and finally (7) how scientific knowledge itself is possible. To be sure, this is a heavy load for one phenomenon to bear in the writings of a major Western philosopher. We must ask then, What is the *essential role* of Angst in Heidegger's thought: Is it truly a "phenomenological window" through which Dasein's Being is self-disclosed? Or is it simply a cognitive tool used by Heidegger to get to the question of Being? Is it both? Or is Angst really Dasein's primordial attunement to Being? This important question must remained unanswered until our concluding chapter.

NOTES

/1/ This description is based upon a photograph taken from within Heidegger's study in the Black Forest. Cf. Walter Biemel, *Martin Heidegger: An Illustrated Study*, trans. by J. L. Mehta (New York and London: Harcourt Brace Jovanovich, 1976), p. 72. Note: in the "Chronology" appendix to this work, we are told that the cottage at Todtnauberg was built in 1922 when Heidegger went to Marburg as an "Associate Professor." Cf. p. 181.

/2/ George Steiner, *Martin Heidegger* (New York: Penguin Books, 1978), p. 74.

/3/ Martin Heidegger, *Sein und Zeit, Erste Hälfte*. In *Jahrbuch fur Philosophie und phenomenologische Forschung*, 8 (Halle: Niemeyer, 1927). We have used here the 6th edition (Tübingen: Neomarius Verlag, 1949), p. 184. Note: the present author is responsible for all translations taken from *Sein und Zeit* as well as from the original Inaugural Lecture "Was ist Metaphysik?" He wishes to acknowledge, however, the assistance of MacQuarrie and Robinson in their English translation of *Being and Time* as well as R. F. C. Hall and Alan Crick in their early translation of "What is Metaphysics?" found in Warner Brock, ed., *Existence and Being* (Gateway edition; Chicago: Henry Regnery Company, 1949), pp. 325–61, and the more recent translation of "Was ist Metaphysik?" by David Krell in *Martin Heidegger: Basic Writings* (New York: Harper & Row, 1977), pp. 95–112. Our attempt at useful translations was undertaken for two purposes: (1) to provide consistency from one of Heidegger's works to the other, and (2) to correct some flagrantly obvious errors found in earlier translations. (We shall follow our translations with *SZ* and give the page number from *Sein und Zeit* in the text. Likewise, translations from "Was ist metaphysik?" will be designated WM.)

/4/ We designate this period as the "early mature period" because by 1926 when *Sein und Zeit: Erstes Haelfte* was completed, Heidegger was already an "Associate Professor" at Marburg, but had not as yet obtained the depth of his later post-1929 period, i.e., the period after the famous *Kehre*.

/5/ Martin Heidegger, *Was ist Metaphysik?* (Bonn: Cohen, 1929), fourth edition with "Postscript" (*Nachwort*) and fifth edition with "Introduction" (*Einleitung*) (Frankfurt: Klostermann, 1943 and 1949). Note: for English translations see note 3 above.

/6/ Martin Heidegger, *Vom Wesen des Grundes*, fourth edition (Frankfurt: Klostermann, 1955), translated into English in a bilingual edition by Terrence Malick, *The Essence of Reasons* (Evanston: Northwestern University Press, 1969).

/7/ Martin Heidegger, *Kant und das Problem der Metaphysik*, second edition (Frankfurt: Klostermann, 1951), translated into English by James S. Churchill, *Kant and the Problem of Metaphysics* (Bloomington and London: Indiana University Press, 1962).

/8/ Heidegger's critics on the concept of Angst are legion, a fact he himself alludes to in the postscript to *Was ist Metaphysik?* A clue to understanding the scope of such criticism is given by Herbert Spiegelberg in Volume I of *The Phenomenological Movement*. After labeling a section head, (5) *Anxiety* (sic) *and Nothingness*, Spiegelberg proceeds to say without further ado: "Few items in Heidegger's philosophy have given rise to more protests and even ridicule than these." Cf. Herbert Spiegelberg, *The Phenomenological Movement*, Volume I (The Hague: Martinus Nijhoff, 1965), p. 331.

/9/ William Richardson saw the essential significance of Heidegger's thinking and questioned Heidegger about it in a letter to him which became the basis for Heidegger's own "Introduction" to Richardson's work, *Heidegger: Through Phenomenology to Thought*. Specifically, Richardson asked Heidegger, "How are we to understand your first experience of the Being question [*Seinsfrage*] in Brentano?" Heidegger responds by offering a full display of the significance of the problem during his early mature period, culminating in Heidegger's repudiation of Husserl's phenomenology. Heidegger says, according to Richardson's translation: "The Being-question, unfolded in *Being and Time*, parted company with this [Husserl's] philosophical position, and that on the basis of what to this day I consider a more faithful adherence to the principle of phenomenology." The reader might observe that this letter was written to Richardson in April of 1962. Cf. William J. Richardson, S. J., *Heidegger: Through Phenomenology to Thought* (The Hague: Martinus Nijhoff, 1963), pp. viii-x.

/10/ Martin Heidegger, "Mein Weg in die Phaenomenologie," *Herman Niemeyer zum Achtzigsten Geburtstag*, April 16, 1963 (privately printed). English translation by Joan Stambaugh, *On Time and Being* (New York: Harper & Row, 1972), pp. 74–82. (Hereafter "Mein Weg.")

/11/ Ibid., p. 74.

/12/ Ibid.

/13/ Ibid., p. 75.

/14/ Bremel, *Heidegger*, p. 181.

/15/ Heidegger, "Mein Weg," p. 75.

/16/ Steiner observes that Heidegger derived at least some of his background material on primordial Angst from Karl Barth's commentary on *The Epistle to the Romans* (1918), which led to Heidegger's examination of Kierkegaard's work on theological Angst. Steiner also observes that other influences on Heidegger's Angst concept include: Rudolf Bultmann's notion of a "theology of crisis" and de-mythologization; studies and lectures on St. Paul, St. Augustine, and Luther; and finally Pascal. Cf. Steiner, *Martin Heidegger*, pp. 73–74. On another occasion it would be important to display the manner of such influences, but the scope of the present essay precludes such an analysis.

/17/ Heidegger, "Mein Weg," p. 75.

/18/ Ibid., p. 76.

/19/ Ibid.

/20/ Ibid.

/21/ Edmund Husserl, *Ideen an einer reinen Phenomenologie und phaenomenologischen Philosophie* (1913). Translated into English by W. R. Boyce Gibson as *Ideas: General Introduction to Pure Phenomenology* (New York: Collier Books, 1962), p. 37.

/22/ Heidegger, "Mein Weg," p. 79.

/23/ Ibid.

/24/ William Dilthey, *Gesammelte Schriften*, III (Stuttgart: Teuber, 1958–61), p. 210, as cited in J. L. Mehta, *Martin Heidegger: The Way and the Vision* (Honolulu: University of Hawaii Press, 1976), p. 13.

/25/ Ibid.

/26/ Ibid., p. 291.

/27/ In *Sein und Zeit* Heidegger generally acknowledges Kierkegaard's influence through several footnotes; this is especially so on page 140 where Angst is explicitly mentioned. Says Heidegger regarding the phenomena of fear and Angst: "The most comprehensive advancement in the analysis of Angst as a phenomenon and certainly most comprehensive in the theological context of a psychological exposition regarding the problem of hereditary sin was accomplished by Søren Kierkegaard."

/28/ Mehta, *Martin Heidegger*, p. 8.

/29/ Michael Wyschogrod, *Kierkegaard and Heidegger: The Ontology of Existence* (New York: Humanities Press, 1969).

/30/ Seidel, in his work on Heidegger and the pre-Socratic thinkers, points out that the origins of Heidegger's thought can be traced back to Parmenides and Heraclitus. Thus they too, along with our six thinkers discussed here, must be included in the list of primary influences on Heidegger's thought. Cf. George J. Seidel, O. S. B., *Martin Heidegger and the Pre-Socratics: An Introduction to His Thought* (Lincoln: The University of Nebraska Press, 1964), p. 2, passim.

/31/ Richardson, *Heidegger*, p. 27.

/32/ This was the promised first section of Part II of *Sein und Zeit*, which was never published.

/33/ Mehta, *Martin Heidegger*, p. 22.

/34/ John Sallis, ed., *Heidegger and the Path of Thinking* (Pittsburgh: Duquesne University Press, 1970), p. 11.

/35/ J. L. Mehta, *The Philosophy of Martin Heidegger* (Varanski, Banaras Hindu University Press, 1967), p. 116, footnote. See note 36 below.

/36/ Heidegger himself points this out in his *Brief über den Humanismus* (*Letter on Humanism*) to Jean Beaufret of December, 1946. In that letter Heidegger states: "Sartre's key proposition about the priority of *existentia* [actuality] over *essentia* [essentiality] does, however, justify using the name 'existentialism' as an appropriate title for a philosophy of this sort. But the basic tenet of of 'existentialism' has nothing at all in common with the statement from *Being and Time*—apart from the fact that in *Being and Time* no statement about the relation of *essentia* and *existentia* can yet be expressed since as yet there is still a question of preparing something cursory." Cf. Martin Heidegger, "Brief ueber den Humanismus," in *Platons Lehre von der Wahrheit* (Bern: Franke, 1947); English translation by Frank A. Capuzzi as "Letter on Humanism," in Krell, *Martin Heidegger: Basic Writings*, p. 209.

/37/ The use of the first person personal pronoun is the phenomenological "I," as is used following Husserl. We shall continue its usage wherever appropriate throughout this work.

/38/ Cf. section 38 of *SZ*, pp. 175–80, passim.

/39/ Heidegger, "Mein Weg," p. 80.

/40/ Heidegger, "Letter on Humanism," p. 208.

/41/ Mehta, *Martin Heidegger*, p. 173.

/42/ *SZ*, p. 178.

/43/ "Factical" generally *means* "contingent," in the sense of finding oneself (*sich befinden*), or, as Versenyi observes, "committed to the mode of Being in which it [Dasein] finds itself, on its own, abandoned to its own limited powers and resources, without having any explanation of the fact (of its being-there) that is here revealed to it." Cf. Laszlo Versenyi, *Heidegger, Being and Truth* (New Haven and London: Yale University Press, 1965), p. 19. Cf. *SZ*, pp. 135–36.

/44/ We coin this expression to display the full structure of *Befindlichkeit* as being attuned to the world and thereby disposed to such an attunement; a factical state which shows itself in our moods.

/45/ *SZ*, section 30, pp. 140–42.

/46/ "*Das Worum die Furch Fürchtet* . . . may alternatively be translated as 'that about which fear is fearful,'" the translation provided by MacQuarrie and Robinson. Cf. *SZ*, p. 141.

/47/ *SZ*, pp. 140–41.

/48/ Ibid., p. 141.

/49/ "*Grauen*," a word meaning: "to have a horror or an aversion to something," is clearly closer to "dread" than is the term "Angst."

/50/ "*Bändigkeit*," a word meaning: "anxiety, uneasiness, apprehension, fear, dread," is clearly an ontic mode of phenomenological and primordial Angst. Hence to glibly interpret Angst as "anxiety" fails to grasp the latter's phenomenological significance from the very outset. Cf. note below.

/51/ It is somewhat curious that Wyschogrod chooses to neglect a comparative analysis of Angst in Kierkegaard and Heidegger, although he does discuss Heidegger's notion of Angst in *Being and Time*. Cf. Wyschogrod, *Kierkegaard and Heidegger*, pp. 106–7.

/52/ *SZ*, p. 189.

/53/ Ibid. Cf. Section 41 through Section 44.

/54/ Ibid., p. 249.

/55/ *SZ*, pp. 250–51.

/56/ However, as Heidegger was later to state explicitly in the 1943 "Postscript" to *What is Metaphysics?*, this does not mean that he is advocating a kind of "heroic philosophy" which goes beyond the basic question with which he is concerned—the question of the meaning of Being. Heidegger is merely pointing out that primordial Angst is systematically and fundamentally misunderstood and interpreted as a psychological anxiety, suffered by those who are "fainthearted."

/57/ This, of course, is another gnostic theme that Heidegger may have borrowed after his 1921 lectures on Gnosticism.

/58/ Later, "ex-istence," to designate Dasein's "standing in the light of Being." Cf. Heidegger, "Letter on Humanism," p. 204.

/59/ Heidegger, "Letter on Humanism," p. 208.

/60/ As was indicated in note 3 above, the present author accepts responsibility for the translations from "Was ist Metaphysik?" Our source for such translations is Martin Heidegger, "Wegmarken," *Gesamtausgabe*, Band 9 (Frankfurt: Klostermann, 1975). The pages given here will correspond to those in the lecture followed by those in the *Gesamtausgabe*, separated by a slashmark. The lecture will be designated *WM* and the *Gesamtausgabe* by *G*.

/61/ This is an extremely important statement, because it implies that man can be changed into *Da-sein only* through the phenomenon of Angst. If so then Angst undergirds the whole of fundamental ontology, carrying with it vast significance beyond what was revealed in *Being and Time*. Heidegger's own words are: "Geschieht in Dasein des Menschen ein solches Gestimmtsein, in dem er vor das Nichts selbst gebracht wird? Dieses Geschehen ist möglich und auch wirklich—wenngleich Selten genug—nur fur Augenblicke in der Grundstimmung der Angst." This we translate as: "Does there occur in the Dasein of Mankind such a kind of attunement in which Dasein is brought face to face with the Nothingness itself? This occurrence is not only possible—it is real—although infrequent enough—for a moment in the primordial mood of Angst" (*WM*, p. 8/*G*, p. 111).

/62/ Martin Heidegger, "Nachwort zu: 'Was ist Metaphysik?'" *Gesamtausgabe*, Band 9 (Frankfurt: Klostermann, 1976). Here we shall designate the postscript passage citings as *Nachwort*. We shall give the original pagination followed by that in the *Gesamtausgabe*, again separated by a slash.

/63/ Mehta, *Martin Heidegger*, p. 7.

/64/ *SZ*, p. 13.

/65/ Ibid.

/66/ Plato, *Timaeus*, 29d, 1–3.

/67/ Caputo, *The Mystical Element of Heidegger's Thought*, p. 223, passim. While Caputo recognizes, of course, that Heidegger rejects the name of "Mystic" as vehemently as he does the name "Existentialist," the parallels between his later thought on Being and the insights of mystical "thought" are simply too obvious to be ignored. Hence, the *Seinsfrage*, too, since it pervades Heidegger's total literary output has the essentials of mysticism. In our view, Caputo simply cops-out when, after 238 pages, he asserts: "The best thing we can do with Heidegger's thought is to leave it uncategorized. . . ." One does not have to be a genius to understand that a philosophical position either has mystical elements or it does not, given virtually any definition of mysticism. In our view, although we cannot develop it fully here, Heidegger *does* have elements of mysticism in his early mature thought; otherwise his discussion of the Nothing and Angst's revelation of it is nothing more than empty words.

CHAPTER IV

TWO EXISTENTIAL THINKERS ON ANGST

> In anguish (*angoisse*) I apprehend myself at once as totally free and as not being able to derive the meaning of the world except as coming from myself.
>
> —Jean Paul Sartre, *Being and Nothingness*

The Question at Hand

On a dark, starless night somewhere in France, a walleyed man looks out of his prison camp window. He listens to an animated interrogation behind the closed door of the office across from his barracks cot. Amid the high-pitched, German-accented questions, he hears a sharp slap, a yelp, and the interrogation continues, punctuated often by muted thuds. The walleyed man jumps at a tug on his shirt sleeve. Startled, he swirls to find a priest. "It's time, Jean-Paul," the priest says softly. The walleyed man looks down the long row of cots. At the far end of the barracks several other priests are waiting for Jean-Paul Sartre to continue his discussion of Heidegger's *Being and Time*. "Yes, I see," says Sartre. Over his shoulder the sounds of violence have stopped. The office door opens and crisp footfalls echo throughout the barracks. Later, outside in the still and starless night, staccato rifle shots drown Sartre's quiet voice as he discusses the role of Angst in Heidegger's thought./1/

Almost a quarter of a century earlier, a young Prussian army chaplain with intense blue eyes stands transfixed in the thundering fire-flashes of the German cannons against the rainy night sky. Numbly he looks down at the blood drenched bodies of his closest friends. Too many lay slowly dying on mud-bespattered stretchers, their screams and death moans weaving a spell of grotesque madness ablaze with cannon fire. Several are already dead, the rain making tiny wet exclamation points on their closed eyelids. One young officer gazes up at the chaplain's insignia, his mangled body broken like a discarded dime-store manikin. A wet cigarette dangles from his lips. One hacking, body-heaving cough and the life-light in his eyes goes out. The chaplain digs his fingernails into his palms and gulps down nausea. The dead, the dying, the red rags on the stretchers where legs and arms should be; the fire-flashes, the cannon's roar, the smoke, the rain and endless mud all swirl in together in a magic madness. With a shudder shaking his being to its very foundations, something changes in the chaplain's blue eyes. Many years later Paul Tillich was to confess that this night at the battle of the Marne ". . . absolutely transformed me. All my friends were among these dying and dead. That night I became an existentialist!" /2/

These two markedly different scenarios convey a common bond running like a broken varicose vein through the first half of the twentieth century: namely, the ultimate frustration, the horror, the terror, the utter absurdity of Dasein's *grande access de folie*, the phenomenon of world-wide war. The hermeneutic implications of war require a philosophical understanding and interpretation beyond even the excellent analysis of J. Glenn Gray's work, *The Warriors: Reflections of Men in Battle.* /3/ Although such a task is beyond our purpose here we do see a direct relationship between the phenomenon of world-war and the rise of *modern* existential thinking of which Sartre and Tillich are articulate representatives. To be sure, there are numerous works of art, history, and existential philosophy that support this interpretation, detailing how the uprootedness of values and vast populations contributed to that intellectual phenomenon known as "existentialism." Our two scenarios grasp graphically the life-styles of two philosophical giants: life-styles that contributed to their respective existential positions, one a-theistic and the other what we call "meta-theistic."

We raise this issue for a singular purpose: to ground phenomenologically a seemingly "psychological" report on the origins of Sartre's and Tillich's respective existential positions. As we see it, Sartre and Tillich revised both their philosophical perspectives and ontological frameworks due to their world-war experiences: in Tillich's case, from a position of Idealism and Neo-Kantianism to modern existential thought; and in Sartre's from transcendental phenomenology to the ethical *praxis* of existential thought. Let us examine how this is so.

Part 1. Sartre on Angst

A Point of Departure

In the two preceding chapters we sought to disclose primordial Angst from the inside out, so to speak. We tried, in other words, to show as a necessary first step the micro-hermeneutical context of Angst's disclosure—without having to undergo any forced or misleading interpretations. With regard to how primordial Angst is revealed in Sartre's thought, we cannot abandon our basic hermeneutical motif despite the double character of the present chapter. To be consistent, therefore, there must be no shortcuts./4/

The Horizons of Influence

To even begin to grasp the phenomenological horizons of Sartre's thought, we must constantly remember that he was first, last, and always a French intellectual whose philosophical roots are deeply embedded in Cartesian dualism. Indeed, Descartes' influence on Sartre is apparent from the very first page of Sartre's first technical phenomenological study, *La Transcendence de l'Ego: Esquisse d'une description phenomenologique*/5/ [*The Transcendence of the Ego: Outline of a Phenomenological Description*]./6/ In this work, Sartre discusses Descartes' *cogito* in the light of two post-Cartesian mainstreams: Kant's critical thought and Husserlian transcendental phenomenology—that is, specifically from Kant's principle of the *cogito*'s universalizability and Husserl's *epoché* of transcendental consciousness. What is Sartre after here? It is the *apodicticity* of the Cartesian *cogito* as the fundamental point of departure for phenomenological studies. Nowhere is this interpretation better confirmed than in the lecture "Existentialism is a Humanism," where Sartre says:

> There can be no other truth to take off from than this: I think; therefore, I exist. There we have the absolute truth of consciousness becoming aware of itself. Every theory which takes man out of the moment in which he becomes aware of himself is, at the very beginning, a theory which confounds truth, for outside the Cartesian *cogito*, all views are only probably, and a doctrine of probability which is not bound to a truth dissolves into thin air. In order to describe the probably, you must have a firm hold on the true. Therefore, before there can be any truth whatsoever,

> there must be an absolute truth; and this one is, simple and easily arrived at; it's on everyone's door step; it's a matter of grasping it directly./7/

Such apodicticity and its implied radical dualism, appears to be Sartre's major appropriation from the Cartesian approach to philosophy. Among Descartes' notions Sartre rejects are both substance and, of course, the notion of God as *causa sui*. Indeed, as Sartre discloses in *Les Mots* (*The Words*), one day, when Sartre was about twelve years old, the Christian God, ". . . tumbled into the blue and disappeared without giving any explanation."/8/ This was, then, the first recogntion of Sartre's avowed atheism.

When both God and substance are abandoned, Sartre is left with the apodicticity of consciousness as the basis of truth. Specifically the "I think," *and* that about which the *cogito* thinks, become an apodictically certain but radical dualism between consciousness per se on the one hand, and what we are conscious *of* on the other. Implied in this point of departure is the fundamental perspective of Cartesian thought: philosophy's basic tool is clear and distinct *intuition* as apodictic certainty rather than the sometimes long and uncertain process of academic argumentation pure and simple. As Marjory Grene has argued, "The *cogito* is not an argument: in the *Meditations*, there is no 'therefore' between 'I think' and 'I am.' I am, I think, Descartes declares, this is true every time I say it."/9/ Sartre seems to agree, and the need for God as *causa sui* is therefore wholly abandoned. Thus, while Sartre abandons Cartesian *theistic* metaphysics in his phenomenological ontology, he, along with Husserl, fully approves of Descartes' method of apodictic intuition as the point of departure for philosophy.

We come, therefore, to the second major influence on Sartre's philosophical development, Edmund Husserl. As was the case with Heidegger, of whom we shall say more presently, Sartre saw the possibility of a new revolution in the philosophical enterprise in Husserl's phenomenological investigations. But Husserl's influence came relatively late in Sartre's development.

In a film interview shot mostly during 1972, entitled *Sartre by Himself*,/10/ Sartre discloses that he was unaware of Husserl as late as 1933. "I didn't know who Husserl was," says Sartre; "he wasn't part of the French cultural tradition. . . ."/11/ But once Husserl did come within Sartre's cognitive sphere, Husserl influenced the young Frenchman immensely. In the interview referred to above, one interviewer, Andre Gorz, asks Sartre: "Where did you get the idea to go to Berlin to study Husserl?" Sartre responds:

> [Raymond] Aron paved the way for me, by introducing me to Husserl's theory. Only in a very cursory fashion, I might add. At the time I read a work by Gurvitch on the intuition of essences in Husserl's work; and I understood that it was very important. And he [Aron] helped me fill out the forms and applications for my trip to Germany; since he had been to the Berlin Institute. So there I was in Berlin, reading Husserl and taking notes on what I read, but without any knowledge of what Husserl was all about, except the smattering I had gleaned from Gurvitch. I didn't even have the vaguest notion about the concept of intentionality. So I set about reading Husserl's *Ideen*./12/

At this point another interviewer, Jean Pouillon, breaks in to ask Sartre: "And in what order did you read Husserl, first the *Ideen*, [*Ideas*], or did you start with the *Logische Untersuchungen*" [*Logical Investigations*]? Sartre replied:

> *Ideen*, and nothing but *Ideen*. For me, you know, who doesn't read very fast, a year was just about right for reading his *Ideen*. I wrote my *Transcendence de l'Ego* in Germany, while I was at the French House there, and I wrote it actually under the direct influence of Husserl; although I must confess that in it I take an anti-Husserl position . . . and after all that long process had taken place, I was absolutely pro-Husserl, at least in certain areas, that is, in the realm of the intentional consciousness, for example; there he really revealed something to me, and it was at that time in Berlin when I made the discovery./13/

From Husserl, Sartre learned at least three major points: First, the pre-reflective *cogito* is the point of departure for philosophical certainty. Second, phenomenology concerns itself with the *cogitationes*, the "objects" of thought through the "intentionality of consciousness" or the target of one's thought. Third, the *epoché*, the phenomenological method of bracketing-out factual notions extraneous to the essence of the phenomenon, is the clue to phenomenological reduction in Husserl's approach. From these perspectives Sartre confidently discussed the manifold classes and types of intentional "objects" that concerned him in

Being and Nothingness—the body, values, the psyche, love, the emotions, and so on—all based upon his comprehension of Husserl's transcendental phenomenology.

On the other hand, Sartre's basic disagreement with Husserl seems to center around one major point: whether there is a "transcendental ego" that governs and ultimately stands behind individual consciousness. Sartre believed that this could not be the case since a transcendental ego would require direct contact with another kind of reality completely different from itself. A two-fold reality structure, he reasoned, would further require yet a third mode of reality that would dialectically capture both the ego and its objects: the so-called *hylé* of Husserl. Sartre believed that this ultimately reduces phenomenology to a transcendental idealism.

What we perceive directly, Sartre reasons, must be seen as material *for* the ego's intentional activity; and the intentional object itself is interpreted as a product of the ego's activity or the "sense data" of consciousness. In Sartre's view this leads to a dangerous idealistic notion that every object of consciousness must relate ultimately to the intentionality of a transcendental ego, a commitment that need not occur if we *reject* the notion of a transcendental ego at the outset. Applying a kind of phenomenological Ockham's razor, Sartre argues that in reality there is no such ego, transcendental or otherwise, that undergirds consciousness: rather "the Ego is the interiority of consciousness when reflected upon by itself."/14/ Thus consciousness has no content: all content is on the side of the objects. So consciousness must be something like pure spontaneity: "a sheer activity of transcending towards objects."/15/ Hence ideas, representations, sense data, etc., are not, according to Sartre, contents *in* consciousness but rather contents *for* consciousness. In short, "consciousness is a great emptiness, a wind blowing towards objects."/16/ For Sartre, therefore, consciousness *is* intentionality.

In *Being and Nothingness*, Sartre expanded this list of charges against Husserl to nine: (1) Husserl suffers from a Berkeleyan idealism by refusing to grant existential status to Being; (2) Husserl is a pure immanationist, and is thus unfaithful to the original aim of phenomenology; (3) Husserl cannot escape the "thing-illusion" by introducing passive *hylé* and the doctrine of sensation into his concept of consciousness; (4) Husserl cannot move to existential dialectics because he remains trapped in mere appearances and purely functional descriptions; (5) Husserl is really a phenomenalist rather than a phenomenologist; (6) Husserl provides merely a caricature of true transcendence, which must get beyond consciousness to world, and beyond the immediate present to the past and the future; (7) Husserl is really a transcendental solipsist in introducing transcendental subjectivity; (8) Husserl cannot deal with the obstructiveness (*coefficient d'adversite*) of our existence which we experience immediately; and finally, (9) Husserl mistakenly thinks that an eidetic phenomenology of essences can lay hold of freedom. From the standpoint of *Being and Nothingness*, the last criticism is clearly the most serious, as we shall see in our discussion of that work./17/

Many of these criticisms became apparent to Sartre only after his encounter with this third great mentor, Martin Heidegger, whom Sartre met personally in 1935./18/ The first mention of Heidegger's name in Sartre's published works occurs in his work of 1939, *Esquisse d'une theorie des emotions* [*Outline of a Theory of the Emotions*]./19/ In this work Sartre ". . . adds Heidegger's ideas to those of Husserl within the same paragraph without implying any essential difference between the two."/20/

Sartre's point of departure, however, which we have seen to be the apodicticity of the *cogito* and therefore of consciousness, placed him in direct opposition with Heidegger's depiction of Dasein as that Being which is already there—thrown into a pregiven world. As we shall soon see, for Sartre Being is radically split into the for-itself (*pour-soi*) or consciousness, and the in-itself (*en-soi*) or the totality of beings within the intentional purview of consciousness. This is a fundamental distinction regarding the nature of Being that cannot be bridged in principle and one which reaffirms the very split Heidegger sought to go behind in *Sein und Zeit*. Nonetheless, Sartre never directly attacks Heidegger regarding Being's unified totality and, in general, he sees Heidegger's solutions to the manifold problems of phenomenology as being superior to Husserl's;/21/ all this despite their fundamentally different views concerning apodicticity and the subject-object implications to which it leads.

More specifically there seem to be six major areas of disagreement between Sartre and Heidegger: (1) Sartre rejected the elimination of the Cartesian and Husserlian notions of consciousness from Heidegger's conception of Dasein. (2) Sartre also objected to the inadequacy of Heidegger's descriptions to capture man as a being whose "projects" bring ontic modifications into the world. (3) Sartre disapproved of Heidegger's insistence that death is man's only authentic project. (4) Sartre believed that Heidegger is completely remiss in failing to see the bodily and sexuality as ontological categories of Dasein. (5) Sartre argued that Heidegger's insistence on the futural mode was primary; this, according to Sartre is one-sided and misleading. (6) Finally, and perhaps most importantly, Sartre utterly rejected Heidegger's attempt to ground the phenomenological conception of the Nothing in the primordial pre-disposition of Angst rather than in the negative element that is the essence of human spontaneity./22/

We may conclude this brief portrayal of the influences on Sartre's emerging career by mentioning their more mature outcomes in Sartre's plan for a synthetic approach to phenomenology. In 1947, he proposed to the *Sociétè francaise de philosophie*,

> a synthesis of Husserl's contemplative and non-dialectical consciousness, which leads us to a contemplation of essences, with the activity of the dialectical, but non-conscious, and hence unfounded project that we find in Heidegger, where we discover that the primary element is transcendence./23/

This is a curious interpretation of Heidegger's thought as being *dialectical* with regard to existence; and we must, of course, agree with Spiegelberg that it is a misinterpretation/24/ if not an outright misrepresentation of Heidegger's thought. But its inclusion in Sartre's proposal is important, because it discloses yet another major influence upon Sartre, namely, the influence of Hegel's dialectical method.

As with Husserl, Sartre did not become aware of Hegel until at least 1933. In the previously cited film interview of 1972, on the coattails of Sartre's admission that Husserl did not come to their early attention, Simone de Beauvoir adds: "We didn't even know who Hegel was!" to which Sartre responds, "That's right, we didn't."/25/

Be that as it may, the dialectical interplay of concepts in *Being and Nothingness*, that is, the specific interplay between Being on the one hand and Nothingness on the other, and the for-itself's desire to become a synthesis, namely, Being-in-itself-of-itself, may be taken as evidence of Hegel's methodological influence upon Sartre that becomes progressively more visible by the time of Sartre's second major work, *Critique de la raison dialectique* (*Critique of Dialectical Reason*)./26/ In *Being and Nothingness*, the whole tenor of Sartre's argument in Section III of the Introduction, entitled "The Dialectical Concept of Nothingness," displays his keen understanding of and appreciation for Hegel's *Treatise on Logic* and the *Phenomenology of Mind* regarding the dialectical concept of Nothingness./27/ What is more, the *en-soi/pour-soi* distinction is, generally speaking, Sartre's expanded re-echo of Hegel's *an-sich* and *für-sich*,/28/ a fact that accords Hegel a special place in Sartre's influences.

As an existential thinker Sartre takes exception to much of Hegel's thought. In *Being and Nothingness*, for example, he takes serious exception to Hegel's conception of Nothingness, noting that: (1) Hegel never got beyond the logical formulation of non-Being, thus failing to see how human reality is encompassed by Nothingness (*BN*, p. 46). (2) Hegel insists that the notions of Being and non-being are logically equal. On the contrary, argues Sartre, it is logical that non-being must be dependent upon Being, that is, non-being must be the negation of an original Being (*BN*, p. 47). (3) Finally, Sartre objects to Hegel's inadvertent bestowal of covert being upon non-being, by granting to both equal ontological status (*BN*, p. 44).

Thus Hegel's influence over Sartre is bi-polar. On the one hand the substantive issues of the *en-soi* and *pour-soi* are reflections of Hegel's *an-sich* and *für-sich*. But far more important for Sartre in the long run was Hegel's dialectical method, a method used by Sartre in varying degrees of rigor throughout *Being and Nothingness*. One may transcend Hegel's categories, but one either employs or chooses not to employ the dialectical method; Sartre is clearly on the first side of this option.

One further potential influence on Sartre's thought may be useful. Sartre may have been influenced by Jules Lequier, the nineteenth-century contemporary of Kierkegaard, and the so-called "father of French existentialism,"/29/ who devoted his entire intellectual life to an analysis of freedom. Sartre may have been introduced to Lequier's thought thanks to Sartre's friend and colleague, Jean Geneve, who in 1936 wrote his doctoral dissertation on Lequier while at the Sorbonne.

Lequier was convinced that the Western philosophical tradition, with the possible exceptions of Aristotle and Fichte, had been covertly opposed to real human freedom, while at the same time pretended to advocate it. He found this to be especially true of the Scholastic philosophers upon whom he lavished a scathing and bitter wit. Jean Wahl in his own work on Lequier/30/ claims that Sartre has found no more fitting formula "for affirming his existentialism" than that devised by Lequier: "FAIRE, non pas *devenir*, mais faire, et, en faisant, SE FAIRE" [TO CREATE, not to *become*, but to create, and in creating, to CREATE ONESELF.]

For us the validity of Wahl's claim of Lequier's influence on Sartre must remain an open question. But right or wrong, Wahl is at least accurate concerning the fitting formula for Sartre's existential thought, a topic to which we will return momentarily. Nonetheless, Lequier is not mentioned in *Being and Nothingness*, nor as far as we know in any other of Sartre's published writings./31/

In conclusion, this truncated sketch in no way pretends to be complete regarding the influences upon Sartre's thought. *Being and Nothingness* is vastly rich in appreciation for and polemic against many influences. The hermeneutical motif of this essay requires, however, our mentioning at least the major influences on Sartre's thought, namely, Descartes, Husserl, Heidegger, Hegel, and possibly Lequier.

With this preliminary hermeneutical sketch completed, we turn now to a general overview of *Being and Nothingness* as the broader hermeneutical context of Sartre's primordial Angst discussion. In this way we hope to prepare the ground for a direct discussion of Angst as *angoisse* in both *Being and Nothingness* and still later in the public lecture entitled "Existentialism is a Humanism."

Some Relevant Elements of Being and Nothingness

Being and Nothingness, Sartre's answer to Heidegger's *Being and Time*,/32/ was originally conceived in 1930. Due to Heidegger's influence, no doubt, Sartre's work was changed in direction and focus during the time that he was a prisoner-of-war in 1940–1941. Consequently, only after his release in 1941 did Sartre write *Being and Nothingness*, publishing it in 1943. These facts may be historically important, but they are not as important hermeneutically as are the circumstances prompting the change in direction of the work. Specifically, in *Sartre by*

Himself, we discover the essential role Heidegger played in this change. The interviewer, Jean Pouillon, observes: "It was when you came back from the prisoner-of-war camp that you wrote *Being and Nothingness*." Sartre responds:

> Yes, because during my stay in the camp I had read Heidegger—I had read him before, but it was then that I really went into his work deeply—and three times a week I used to explain to my priest friends Heidegger's philosophy. And that, plus my own personal thoughts—which actually were a continuation of *Psyche* [an unfinished and unpublished work], a work I had written some time before but which at that point in my thinking was influenced by Heidegger—gave me more or less the elements I needed to put into writing. And my notebooks—which were subsequently lost—were full of observations which later found their way into *Being and Nothingness*./33/

It would appear from this passage that at root this *chef-d'oeuvre* of French phenomenological ontology is basically an ethical, *praxis*-oriented, interpretation of Heidegger's *Being and Time*. Sartre's work stems, as we have seen above, from the standpoint of the Cartesian dualism, the Husserlian phenomenology, and the Hegelian dialectics that we discussed in the previous section.

The Introduction to *Being and Nothingness* is by far the most opaque part of the text. Very broadly speaking, it is concerned with an analysis of the *cogito* as being pre-reflective and with why the pre-reflective *cogito* must be the apodictic point of departure for phenomenological ontology. Sartre contrasts and compares his own position with Idealism and Realism, and explains his theory of the "trans-phenomenality of Being": a doctrine maintaining that ". . . the object of consciousness is always outside and transcendent, that there is forever a resistance, a limit offered to consciousness, an external something which must be taken into consideration."/34/ Moreover, Sartre presents in this section his distinction between the *pour-soi* as Being-for-itself and the *en-soi* as Being-in-itself.

Part I, entitled "the Problem of Nothingness," is concerned with how the *pour-soi* and the *en-soi* as opposites can both participate in Being. Sartre, following Heidegger's definition of Dasein as a questioning entity,/35/ argues that the *pour-soi* can question only on the condition that at the root of consciousness there exists a Nothingness: "Nothingness lies coiled in the heart of being—like a worm" (*BN*, p. 56). Thus, the central focus of part one is twofold: first at the center of his being man is a Nothingness; and second, as a Nothingness man can bring Nothingness into Being. We learn that for Sartre the Nothingness is disclosed to us through primordial Angst, which according to Sartre is best described as phenomenological "anguish." His term, of course, is *angoisse*. Man attempts to flee from anguish, and in doing so he plunges into "bad faith" (*mauvaise foi*)./36/ Through such bad faith, Sartre's equivalent for Heidegger's "inauthenticity," we discover the basic difference between the being of man (as *pour-soi* or consciousness) and the being of the *en-soi* or Being-in-itself (that which is not consciousness). The difference is that man is a being "who is what he is not and who is not what he is."/37/ Man chooses himself, in other words, existing *toward* what he will become in the future. Bad faith is man's attempt to capture his own essence: to say that he *is* a philosopher, an executive, or some other static being. Such bad faith robs humanity of its necessary essence, namely, its freedom as translucent consciousness.

Essential to this discussion is Sartre's intrepretation of the Nothingness. To grasp the meaning of the Nothingness, Sartre tells us, we must not grant to it any covert ontological status, as did Heidegger (in Sartre's view) in his statement: "das Nichts selbst nichtet."/38/ Rather, for Sartre, Nothingness ". . . ne se neantise pas, le Neant 'est neantise'" (*BN*, p. 58). This can only be translated as "Nothingness does not nihilate itself, [but rather] it is nihilated." There is, then, no active existence to Nothingness: it is pure negation—the condition of all verbal as well as experienced negativity. To ground this interpretation, Sartre raises two fundamental questions: (1) What is the relationship we have to the world called "Being-in-the-world?" (2) What must man and the world *be* in order for a relation between them to be possible? (*BN*, p. 34).

With regard to the first of these questions, Sartre agrees with Heidegger: man is a questioning being, a being which as a question asker stands ". . . before a being which we are questioning" (*BN*, p. 35). Now at bottom, argues Sartre, each answer must be either affirmative or negative. Sartre anticipates any objections to this basic distinction by noting:

> There are questions which on the surface do not permit a negative reply—like, for example, the one which we put earlier, "What does this attitude reveal to us?" But actually we see that it is always possible with questions

> of this type to reply, "Nothing" or "Nobody" or "Never." Thus at the moment when I ask, "Is there any conduct which can reveal to me the relation of man with the world?" I admit on principle the possibility of a negative reply such as, "No, such a conduct does not exist." This means that we admit to being faced with transcendent fact of the non-existence of such conduct. (*BN*, p. 35)

Sartre's point here is that there must exist the permanent possibility of a negative reply to any question. For example, in the very simple question "How are you?" there is the negative possibility of a reply such as "I feel terrible!" Here, the "terrible" represents the negativity that stands *behind* the feeling as well as the response, as that to which the response refers, namely, a "Nothingness as its origin and foundation" (*BN*, p. 56). Hence, man's relationship to the world, which we call Being-in-the-world, is one of *questioning* that which is in-the-world in such a manner as to anticipate either a positive or negative reply.

With respect to the second question, "What must man and the world be in order for there to be a relation between them," Sartre responds that the polarities between questioner-questioned determine the character of man (as questioner) and the world (as questioned). As such, the two metaquestions asked above display the interdependence of man and world. Thus the answer to the first inquiry is simultaneously the answer to the second: man is *réalité humaine*, Sartre's equivalent of Heidegger's Dasein (as a questioner), a questioner standing in the transcendental totality of beings in the referential context known as "world." What distinguishes man from the world is his *ability* to question, as well as the more abstract but important possibility of negation in any reply to questions he poses.

In the context of Sartre's analysis of Heidegger's interpretation of Nothingness, the object of our search, "*angoisse*," first appears. Sartre notes that according to Heidegger,

> There exist . . . numerous attitudes of "human reality" which imply a "comprehension" of Nothingness: hate, prohibitions, regret, *etc*. For Dasein there is even a permanent possibility of finding oneself face to face with Nothingness and discovering it as a phenomenon: this possibility is Anguish [*angoisse*]. (*BN*, p. 50)

This passage is important in our view because it establishes as fact that Sartre saw *angoisse* as equivalent to the primordial

Angst of which Heidegger speaks in both *Being and Time* and in *What is Metaphysics?* As will become apparent, however, the *role* of primordial Angst for Sartre is vastly different than that of Heidegger. But the relationship of identity was apparently presupposed by Sartre; and that, of course, is a point of both micro- *and* macro-hermeneutical significance.

In Sartre's section entitled "The Origin of Nothingness" we come to the epicenter of Sartre's interpretation of Nothingness. Here Sartre raises specifically a vastly important question: ". . . where does Nothingness come from?" (*BN*, p. 56). Sartre argues that Nothingness cannot "nihilate itself" as Heidegger thought, because that would make the Nothingness an actor upon itself. To be such an actor would require at least some status as a being. This is, of course, a contradiction since Nothingness cannot at once be and not-be. Hence, Sartre coins a new term that avoids Heidegger's famous "contradiction": *das Nichts selbst nichtet*. Sartre suggests that Nothingness "is made-to-be" (*BN*, p. 57). That is, Nothingness is nihilated by Being and is thus *acted-upon*.

If this is so, then we must raise this question: what kind of being is so constituted that it *can* bring Nothingness into the world? Such a being, says Sartre: (1) cannot be *en-soi*, since that is a manifest contradiction; (2) cannot be passive in relation to Nothingness; and (3) cannot *produce* Nothingness. Rather, this being ". . . must be its own nothingness" (*BN*, p. 58). *Such a being is man*; man seen as *réalité humaine*: a questioner who has the permanent possibility of Nothingness residing in the core of his Being-in-the-world.

This gives rise then to a second question: "What must man be in his being in order that through him Nothingness may come into being?" (*BN*, p. 59). Sartre's answer: *réalité humaine*, precisely because it is a questioning being that can modify Being through the process of questioning. In posing the question, man steps out of Being to retire "beyond a nothingness" (*BN*, p. 69).

The name given to his possibility to secrete an isolating Nothingness into Being is, of course, "freedom." "Human freedom," says Sartre, ". . . precedes essence in man and makes it possible; the essence of the human being is suspended in his freedom. What we call freedom is impossible to distinguish from the *being* of 'human reality'" (*BN*, p. 60).

Sartre then displays how freedom is intimately bound

together with Nothingness, arguing that freedom is what makes the Nothingness possible, for *réalité humaine*. How is this so? First, Sartre points out, *réalité humaine* can detach itself from the world through questioning and doubting.

Freedom provides to man the possibility of such self-detachment. This is a notion analogous to Descartes' doubt, Husserl's *epoché*, Hegel's mediation, and Heidegger's transcendence.

What interests Sartre is the temporal dimension of freedom since remembering, questioning, and doubting, for example, are modes of behavior. In remembering, for example, there exists a distinct separation, a nihilation, between present consciousness and past consciousness, a separation experienced directly and apodictically. Thus says Sartre, "what separates prior from subsequent is exactly *nothing*" (*BN*, p. 64). The Nothingness allows man to deny all or part of the world, the condition necessary for transcending, and likewise for understanding and interpreting, the world.

With regard to the temporal dimension of experience in general Sartre's basic point is: "It is necessary then that conscious being constitute itself in relation to its past as separated by this past by a nothingness" (*BN*, p. 64). This Nothingness structures consciousness, for ". . . consciousness continually experiences itself as the nihilation of its past being" (*BN*, p. 64). The term for this process of nihilation is again "freedom." Indeed, consciousness exists only as consciousness of nihilation, according to Sartre.

Having thus displayed the identity of nihilation, nihilating consciousness, and freedom, Sartre raises a question of basic concern to our own work here:

> If the nihilating consciousness exists only as consciousness of nihilation, we ought to be able to define and describe a constant mode of consciousness, present *qua* consciousness, which would be a consciousness of nihilation. Does this consciousness exist? Behold, a new question has been raised here: if freedom is the being of consciousness, consciousness ought to exist as consciousness of freedom. What form does this consciousness of freedom assume? In freedom the human being *is* his own past (as also his own future) in the form of nihilation. If our analysis has not led us astray, there ought to exist for the human being, in so far as he is conscious of being, a certain mode of standing opposite his past and

> his future, as being both his past and his future and as not being them. We shall be able to furnish an immediate reply to his question; it is in anguish (*angoisse*) that man gets the consciousness of his freedom, or if you prefer, anguish is the mode of being of freedom as consciousness of being; it is in anguish that freedom is, in its being, a question for itself. (*BN*, p. 65)

With this passage we have arrived at Sartre's specific and detailed interpretation of primordial Angst, as the unique phenomenon wherein man, as *réalité humaine* or Dasein, ". . . gets the consciousness of his human freedom." Let us now examine his description and interpretation in closer detail.

Primordial Angst in Being and Nothingness

Sartre begins his analysis and description of Angst by pointing out that the depictions of Angst by Kierkegaard and Heidegger seem compatible. This is Sartre's point of departure regarding Angst:

> Kierkegaard describing anguish (*angoisse*) in the face of what one lacks characterizes it as anguish in the face of freedom. Heidegger, whom we know to have been greatly influenced by Kierkegaard, considers anguish instead as the apprehension of Nothingness. These two descriptions of anguish do not appear to us contradictory; on the contrary the one implies the other. (*BN*, p. 65)

For Sartre *angoisse*, like Janus of Roman mythology, wears two distinct faces: the first looks toward *réalité humaine*'s future and the second looks towards its past. We shall discuss these separately.

Regarding the Janus face of Angst that looks to the future, the descriptive paradigm is once again the abyss metaphor. Sartre begins this discussion by observing that Kierkegaard was right: "anguish is distinguished from fear in that fear is fear of beings in the world whereas anguish is anguish before myself" (*BN*, p. 65). Sartre proves his point through a phenomenological analysis of vertigo, an expanded revision of subjective Angst in the face of the abyss as was discussed by Kierkegaard. In Sartre's interpretation, vertigo is not a fear of falling into the abyss; rather it is my "fear"/39/ of throwing myself into the abyss as a deliberate act of choice. *Angoisse*, then, is not a fear of some threatening thing. Rather says Sartre, *angoisse* becomes visible as man's fear of fear. For man faces *angoisse* when he, like the recruit who reports for active duty, is more afraid of being afraid than of death itself.

But the phenomenon of *angoisse* is highly complex: if a threatening situation is apprehended from the standpoint of how I expect to handle a dangerous situation, then what I experience is fear. But if the same situation is apprehended from the viewpoint of *how it will affect my self if I mishandle the situation*, then what I am really experiencing is *angoisse*. The important difference between these two is that in the latter *réalité humaine* has an internal apprehension of the "Self" in the face of what threatens; but in the former situation, what threatens *réalité humaine* is ". . . a possibility of my life being changed from without" (*BN*, p. 66). Yet, these phenomena, says Sartre, are intimately bound together in the constant internal switching: experiencing fear then switching to our apprehensive consciousness of fear and vice-versa. In such a switching, when one becomes self-conscious of one's fear, one then experiences the *angoisse* of reflective apprehension of the Self caught in a web of fear-provoking circumstances. Sartre offers the following example of switching: when I lose all my money in a stock market crash, I may be afraid of impending poverty. But a moment later, when I wring my hands and say to myself: "What am I going to do? What am I going to do?" I have imported my *Self* into the poverty threatening situation—hence changing my fear of poverty into a *reflective apprehension of my Self* as that which gets threatened. In this reflective grasp, my feeling switches from fear to *angoisse*.

However, continues Sartre, there are some circumstances where *angoisse* is not mixed with fear, such as, for example, when I am charged with an important mission well beyond my capabilities. In this case I may experience *angoisse* but not fear at all with regard to the possibility of failure; for me failure is not at stake, rather it is how well I will rise to the challenge of the mission.

Hence in these examples Angst as *angoisse* is revealed in different ways. As vertigo in the face of the abyss, anguish is revealed first in fear—the precipice is to be avoided, as in the stones upon which I could slip, or the crumbling turf that may give way beneath me. These circumstances are threats to be avoided but are external possibilities which can *act upon* me at any time. Thus they are possibilities to be feared. Since such possibilities are precisely mine as I stand at the face of the abyss, they exist for me only when I gape transfixed into the beckoning fathomlessness. At that moment I discover that it is I myself who sustain my potential being; that is, whether I choose to plunge into the abyss or not. The "not" here is important. The "not" shows that such possibilities *can* be negated—as, say, by my calmly stepping back from the abyss. This further shows us that such possibilities are neither external nor are they to be feared! Indeed, argues Sartre, *as* my own potential, each internal possibility belongs only to me. Hence it is I who am the ultimate and the permanent source of their negation. How is this so? Standing there at the face of the abyss *I must necessarily* negate the fear-producing possibilities by choosing to make one, and only one possibility actual. Yet, cautions Sartre, there is nothing *causal* about this choosing process if by "causal" we mean something like "psychologically determined." Rather, at the face of the abyss I am in *angoisse*

> . . . precisely because any conduct on my part is only *possible*, and this means that while constituting the totality of motives *for* pushing away that situation, I at the same moment apprehend these motives as not sufficiently effective. . . . At the very moment when I apprehend my being as *horror* of the precipice, I am conscious of that horror as *not determinant* in relation to my possible conduct. In one sense that horror calls for prudent conduct, and it is in itself a pre-outline of that conduct; in another sense, it posits the final developments of that conduct only as possible, precisely because I do not apprehend it as the cause of these final developments but as need, appeal, *etc.* (*BN*, p. 68)

To avoid fear, therefore, I quickly take refuge in reflection. But every form of refuge has its price: reflection offers only a *possible* future. This means precisely that *nothing* can compel me to adopt one possibility over another; neither my background, upbringing, education. Rather ". . . in establishing a certain conduct as a possibility and precisely because it is my possibility, I am aware that *nothing* can compel me to adopt that conduct" (*BN*, p. 68). Nonetheless, I am already projected into my future since my possibilities are potential future actualities. What all this means, then, is that *angoisse* ". . . is precisely my consciousness of being my own future, in the mode of not-being" (*BN*, p. 68). *Nothing* prevents me from throwing myself headlong into the abyss, just as *nothing* compels me to save myself. That decision (choice) must be made by a future Self: "The decisive conduct," says Sartre, "will emanate from a self which I am not yet" (*BN*, p. 69).

With these clues to guide his reader, Sartre describes *humaine réalité* in the face of the abyss:

> I approach the precipice, and my security is searching for myself in my very depths. In terms of this moment I play with my possibilities. My eyes, running over the abyss from top to bottom, imitate the possible fall and realize it symbolically; at the same time suicide, from the fact that it becomes a *possibility* possible for *me*, now causes to appear possible motives for adopting it (suicide would cause anguish to cease). Fortunately these motives in their turn, from the sole fact that they are motives of a possibility, present themselves as ineffective, as non-determinant; they can no more *produce* the suicide than my horror of the fall can *determine me* to avoid it. It is this counter-anguish which generally puts an end to anguish by transmitting it into indecision. Indecision in its turn calls for decision. I abruptly put myself at a distance from the edge of the precipice and resume my way. (*BN*, p. 69)

We have attempted to fully detail Sartre's thinking on futural *angoisse*, because his paradigm is our own, namely, Angst in the face of the abyss. Having worked our way through Sartre's analysis, description, and interpretation in some detail, we may now proceed to review the other features of Angst in this first part of *Being and Nothingness.*

There is next the second Janus face of *angoisse*, the one that looks toward the past. Here the descriptive paradigm is that of a reformed compulsive gambler who, as he approaches the gambling table, experiences profound *angoisse* when he suddenly sees all of his former resolutions not to gamble melt away. *Angoisse* shows our gambler that yesterday's resolutions about "no more gambling" mean absolutely *nothing*. Why?: because what one is *now* is not what one was when the resolution was made yesterday. More specifically, a Nothingness has slipped in between his past resolutions (choices) and his present memory of those resolutions. Says Sartre:

> It seemed to me that I had established a *real* barrier between gambling and myself, and now I suddenly perceive that my former understanding of the situation is

> no more than a memory of an idea, a memory of a feeling. In order for it to come to my aid once more, I must remake it *ex nihilo* and freely. (*BN*, p. 70)

Moreover, this need to recreate the memory for the present displays a magical attempt of past resolutions to free our gambler from *angoisse*. Sartre continues:

> After having patiently built up barriers and walls, after enclosing myself in a magic circle of resolution, I perceive with anguish (*angoisse*) that *nothing* prevents me from gambling. The anguish *is me* since by the very fact of taking my position in existence as consciousness of being, I make myself *not to be* the past of good resolutions *which I am*. (*BN*, p. 70)

Angoisse in the face of the past, therefore, also reveals to *réalité humaine* the absolute freedom of human consciousness. From absolute freedom, and the *angoisse* in which freedom shows itself, there is no final escape, no exit, ever.

Angoisse displays the ultimate freedom of *humaine réalité*: a freedom defined as ". . . a constantly renewed obligation to remake the *Self* which designates a human being" (*BN*, p. 72). This Self is, for Sartre, the essence of being human, an essence grounded in the temporal category of "what-has-been." Thus for Sartre as well as for Hegel and Heidegger, the essence of man is grounded in his past, or his "what-has-been." So *angoisse* appears as an apprehension of this essential Self as it is detached from its "now." I am, in other words, always one step ahead of my self (my essence); I stand-out, I ex-ist, from that essence, choosing my own existence as a result of the Nothingness that stands between that essence and my actualized possibilities. Hence, for Sartre, *angoisse* ". . . is the recognition of a possibility as *my* possibility; that is, it is constituted when consciousness sees itself cut off from its essence by nothingness or separated from the future by its very freedom" (*BN*, p. 73). This means, then, that I am separated by a nihilating *nothing* from my previous choices as well as those standing before me in the future.

The Nothingness that has eluded us now shows itself precisely as our absolute freedom, viewed from the standpoint of human reality. But what we authentically apprehend is *angoisse* in the face of this freedom; for freedom is bounded by *nothing*. "In anguish (*angoisse*) freedom is anguished before itself inasmuch as it is instigated and bound by *nothing*" (*BN*, p. 73). This means, of course, that *angoisse* is precisely a permanent structure of human being. How then do we account for its rarity in our ordinary experience?

First, says Sartre, we must recognize that *angoisse* is the apprehension of a specific possibility as my own possibility; "that is, it is constituted when consciousness sees itself cut from its essence by nothingness or separated from the future by its very freedom" (*BN*, p. 73). In other words, I make an intentional *appointment* with myself on the other side of a future possibility—a future that is separated from my essence (*my* "what-has-been") by a Nothingness. From this perspective *angoisse* is seen as ". . . the fear of not finding myself at the appointment, of no longer wishing to bring myself there" (*BN*, p. 73).

Second, *angoisse* is rare because I find myself engaged in the everyday world; and moreover, engaged in such a way so as to actively prevent my apprehension of possibilities as pure possibilities. "The consciousness of man *in action*," says Sartre, "is non-reflective consciousness" (*BN*, p. 74). Hence, *angoisse* is not visible in the immediacy of experience. But when such non-reflective consciousness becomes reflected upon, as for example, when the writing of this present sentence is seen as part of the yet-to-be-completed essay, consciousness is directed toward the total project as a *finished project*. Precisely then does *angoisse* show itself. For in seeing the work as a completed totality, I recognize that *nothing* prevents me from giving up the work, of never completing it, and of abandoning it forever. Thus when I fully apprehend the negative possibility of *not completing* my fundamental projects, projects vastly important to me, *angoisse* reaches out and seizes me as from the dark side of a dream; for it is I myself who must constantly choose to make actual that which for the moment is merely possible.

Third, *angoisse* is rare because it is not visible in terms of everyday morality. Ethical *angoisse* is revealed when I consider my Self in relation to the values *I create*. Hence, my values are contingent upon my freedom. And, says Sartre, "my freedom is anguished at being the foundation of values while itself without foundation" (*BN*, p. 76). But on the everyday level of morality, I never question my values. The immediacy of experiencing the world as well as the exigency of my dwelling in a value-laden world where I constantly must engage myself ". . . cause values to spring up like partridges" (*BN*, p. 76). In the world of my everyday concerns, then, I discover myself as a being engaged in a situation, in an enterprise, in a project. The meaning of my ultimate project, which is my life history, is determined by me alone. *I choose myself!* I project, moreover, this Self-made choice upon the world so that ". . . values, appeals, expectations and in general a world . . ." (*BN*, p. 77) appear to me as the transcendental value horizon of my individual projects, as the referential meaning-totality of my individual experience, and as the logical signification and justification of my individual world. But as regards the ontic world, the world of mundane life, Sartre says:

> For the rest, there exist concretely alarm clocks, signboards, tax forms, policemen, so many guards against anguish (*angoisse*). But as soon as the enterprise is held at a distance from me, as soon as I am referred to myself because I must await myself in the future, then I discover myself suddenly as the one who gives its meaning to the alarm clock, the one who by a signboard forbids himself to walk on a flower bed or on the lawn, the one who decides the interest of the book which he is writing, the one finally who makes the values exist in order to determine his action by their demands. (*BN*, p. 77)

And what then is the point of this description in regard to *angoisse*? Says Sartre:

> I emerge alone and in anguish (*angoisse*) confronting the unique and original project which constitutes my being; all the barriers all the guard rails collapse, nihilated by the consciousness of my freedom. I do not have nor can I have recourse to any value against the fact that it is I who sustain values in being. (*BN*, p. 77)

With these three points concerning the rarity of Angst's visibility completed, Sartre concludes his discussion of *angoisse* with this final definition: "Anguish (*angoisse*) then is the reflective apprehension of freedom by itself" (*BN*, p. 78). This is an important addition to the literature on Angst because it defines *angoisse* as a *state of reflective consciousness: that is, a state consciously given to the Self above the prereflective level.* Thus Angst, interpreted an *angoisse*, is a Self-conscious reflection of pre-reflective freedom; a freedom that springs from the Nothingness that lies ". . . coiled in the heart of being—like a worm" (*BN*, p. 56).

We come then to flight in the face of *angoisse*. For Sartre *angoisse* is the reflective apprehension of absolute freedom, as indeed we have seen. Thus I myself am the alpha point of the world's meaning and so too of the values I create within this referential context called world. Yet I flee from the terrifying responsibility entailed in such ultimate freedom. In particular, flight is particularly displayed in the "psychological determinism" that covers over my essential Angst; for determinism denies ". . . that transcendence of human reality which makes it [i.e., human reality] emerge in anguish (*angoisse*) beyond its own essence" (*BN*, p. 79). So psychological determinism emerges as a foil whereby *angoisse* is easily dealt with and thereby rendered harmless. It is, therefore, a merely personally satisfying excuse for absolute freedom, one that robs *angoisse* of its primordial sting. As such it leads me into distraction from the primordial power of *angoisse* in the face of the future. But such flight also applies to *angoisse* in the face of the past. Specifically, to avoid or cover over such *angoisse* I assert (and thereby try to convince myself) that I *am* my essence ". . . in the mode of being of the in-itself" (*BN*, p. 81). Refusing to consider my essences as being historically constituted, I attempt to flee from my essential transcendence by interpreting freedom as a property inherent in my individual ontic self as though I am a little god who must possess freedom as a metaphysical virtue. This reassures me, allowing me to bask in the autonomous smugness of a universal order to which I as an ontic self, belong. But to do this is a fiction that obscures completely the true nature of my essence: a being wholly grounded in what-has-been, a being standing out in the present through Nothingness, on the way to actualizing future possibilities.

But can such attempts to flee from *angoisse* succeed? Sartre says without reservation on this score: "It is certain that we cannot overcome anguish (*angoisse*), for we *are anguish*" (*BN*, p. 82). To be sure, I can cover it over and hide it by inauthentic means; but in doing so I must first admit that *angoisse* is there originally, a powerful phenomenon to be hidden from me. Thus, *angoisse* is finally revealed as the existential truth that I must hide from my ontic Self. If so, then *angoisse* must be acknowledged as hiding or covering-over its profound power over me. This is the origin of psychological determinism, that theory by which I overcome my primordial Angst.

Sartre concludes his *angoisse* descriptions and interpretations by offering the following observation on "bad faith" as an attempt to flee primordial Angst's power:

> . . . to flee anguish [*angoisse*] and to be in anguish cannot be exactly the same thing. If I am my anguish in order to flee it, that presupposes that I can decenter myself in relation to what I am, that I can be anguish in the form of "not-being-it," that I can dispose of a nihilating power at the heart of anguish itself. This nihilating power nihilates anguish in so far as I flee it and nihilates itself in so far as *I am anguish in order to flee it*. This attitude is what we call *bad faith*. There is then no question of expelling anguish from consciousness nor of constituting it in an unconscious psychic phenomenon; very simply I can make myself guilty of bad faith while apprehending the anguish which I am, and this bad faith, intended to fill up the nothingness which I *am* in relation to myself, precisely implies the nothingness which it presupposes. (*BN*, p. 83)

We turn now to one further discussion of *angoisse* in Sartre's thought. For Sartre, as did Heidegger before him, presents his final statement on this complex phenomenon in the form of a lecture. For Sartre this was "L'existentialisme est un humanisme," which Walter Kaufmann believes is Sartre's "best known work" in the English-speaking world./40/

Primordial Angst in "Existentialism is a Humanism"

The word "Existentialism" never appears in *Being and Nothingness*. It first appears within Sartre's works in his concise *Mise au point* (Clarification) in reply to Communist attacks against *Being and Nothingness* in the paper *Action*, on December 20, 1944./41/ By 1946 these attacks were joined against Sartre by French Catholics who took exception to his atheism. Hence, largely as an answer to both the Catholics and the communists, Sartre delivered an *apologia* of "existentialism" during 1946. While he was to later publicly state his regret in allowing the lecture to be published, Sartre's impact on the world was profound indeed—the lecture inadvertently became the *locus classicus* of the philosophy of "existentialism."/42/

Sartre states his atheism unequivocally in the course of the lecture by saying:

> Atheistic existentialism, of which I am a representative, declares with great consistency that if God does exist there is at least one being whose existence comes before its essence, a being which exists before it can be defined by any conception of it. That being is man, or, as Heidegger has it, the human reality. What do we mean by saying that existence precedes essence? We mean that man first of all exists, encounters himself, surges up in the world—and defines himself afterwards. If man as the existentialist sees him is not definable, it is because to begin with he is nothing. He will not be anything until later, and then he will be what he makes of himself. Thus, there is no human nature, because there is no God to have a conception of it. Man simply is. . . . Man is nothing else but that which he makes of himself. That is the first principle of existentialism./43/

In this passage above, Sartre reaffirms the basic position of *Being and Nothingness*—Self choice. "Man is responsible for what he is."/44/ Thus "existentialism" is a humanism, because all values are human values. As a humanism, "existentialism" has at its own center the belief that man not only chooses himself but in so choosing, he chooses at the same time for all men; since ". . . of all the actions a man may take in order to create himself as he wills to be, there is not one which is not creative, at the same time, *of an image of man such as he believes he ought to be*."/45/ (Italics added.) Hence, my individual choice reflects my values as universal values within the context of my historical epoch. I am in this sense as responsible for all men as I am for choosing myself; my "existentialism" is, in short, a humanism and an ethical humanism at that.

But what does *angoisse* have to do with all of this? Sartre responds that I am always in *angoisse* precisely because of the universality of my self-choices as reflections of universal values. My responsibility, therefore, is the ground of my *angoisse*. Even those persons who never display *angoisse*, who cover it over or who flee from it, are constantly in *angoisse*. From *angoisse* there is literally no exit; for my very need to disguise *angoisse* vividly displays the urgent and primordial power it has over me.

Sartre cites Kierkegaard's example of Abraham's *angoisse* at the angel's command to sacrifice Isaac, drawing attention to the fact that anyone in Abraham's place would wonder if this angel is really an angel and, more to the point, ". . . whether I am really Abraham. Where are the proofs?"/46/ In a similar fashion, says

Sartre, who can prove to me that I am individually the proper person to impose my individual conceptions of man upon mankind as a whole? There is no proof here either, yet I must at each and every moment perform exemplary actions *as though I am fully convinced* of the essential rightness of imposing my own values upon others in a universal way. Sartre observes regarding this matter:

> Everything happens to every man as though the whole human race had its eyes fixed upon what he is doing and regulated its conduct accordingly. So every man ought to say, "Am I really a man who has the right to act in such a manner that humanity regulates itself by what I do." If a man does not say that, he is dissembling his anguish (*angoisse*)./47/

Such *angoisse*, then, is not the kind of phenomenon that leads to quietism or inaction: its very essence is *bound* to action and the responsibility I must take for my actions. Using the descriptive paradigm of the military man, Sartre points out that the responsibility of the commander to send his men into battle—where clearly some must die—is a personal choice made in the ultimate alone-ness of responsible decision. But even if he is acting under orders from his superiors, the commander must *interpret* such orders. Upon his, and ultimately *only* his, personal interpretation depend the lives of his men. All leaders, claims Sartre, know the *angoisse* associated with such decisions. Yet *angoisse* does not prevent such men from making such decisions; indeed it makes their decisions possible; ". . . for the action presupposes that there is a plurality of possibilities, and in choosing one of these, they realize that it has value only because it is chosen."/48/ This responsibility as my own responsibility cannot be passed over to another and is the root of the essential Angst revealed in this lecture. Sartre concludes his discussion of Angst by observing: ". . . it is anguish (*angoisse*) of that kind which existentialism describes, and moreover, as we shall see, makes explicit through direct responsibility towards other men who are concerned. Far from being a screen which could separate us from action, it is a condition of action itself."/49/ With this observation Sartre draws to a close his discussions of the place and role of *angoisse* in the lecture. This completes as well our own interpretation of Angst in *Being and Nothingness* and "Existentialism is a Humanism." In the following section we must raise several significant issues that require further discussion and clarification in Chapter V.

Preliminary Conclusions Regarding Sartre and Angoisse

We must draw to a close this part of the present chapter by again raising the question of what we have learned regarding Sartre's understanding and interpretation of Angst as *angoisse*. Bearing in mind that any answer must remain provisional until our fifth and final chapter, we suggest six points require further discussion. These are: (1) Sartre's attempt to unify Kierkegaard's and Heidegger's Angst interpretations results in returning Angst to the province of phenomenological psychology. (2) Sartre's analysis of absolute freedom and the Nothingness within *réalité humaine* implies that Angst is wholly immanent in man. (3) Sartre's analysis of Angst and values creation clearly goes beyond the purely ontological concerns of Heidegger, returning us to Kierkegaard's *ethical* concerns but now on a purely atheistic level. (4) Sartre's analysis of Angst is essentially negative in two senses: (a) "ontologically" negative, because it is the source of introducing Nothingness into Being; and (b) "psychologically" negative because *angoisse* "is the Self already being afraid" (*BN*, p. 65). (5) Sartre's use of *angoisse* as an anthropological nexus between Being and Nothingness *within* man also serves as the ontological nexus between the *pour-soi* and the *en-soi*. (6) Sartre's interpretation of *angoisse* as that from which man seeks escape in "bad faith" provides a launching point for almost all dimensions of Sartre's existential analysis of man: Angst may be said, therefore, to be a major foundational doctrine of Sartre's "existentialist" philosophy.

Here, as well, these accomplishments and shifts in the mode of Angst's revelation give rise to several issues and concerns demanding further clarification before the ultimate value of Sartre's contribution can be judged. For as we shall see, these concerns are rooted in the primary assumptions of Sartre's "existentialist" doctrine. Thus criticisms go well beyond any discussion of *angoisse*.

1. As we have seen, Sartre's account of the psychological phenomenon of *angoisse* is grounded in what he regards as the ultimate and apodictically certain distinction between the *pour-soi* and the *en-soi*. Hence we must *at the outset accept this*

interpretation as being a more adequate picture of reality than Heidegger's attempt to get beyond such dualistic interpretations of Being. Indeed, if we cannot accept Sartre's interpretation, then his analysis of *angoisse*, as one fully contingent upon his fundamental dualism, likewise cannot be accepted as either valid or sound.

2. This gives rise to another question regarding the "existentialist" underpinnings of *angoisse*. By capturing Angst in a web of anthropological immanence, Sartre is guilty of constructing a vast and richly subtle misunderstanding of the goals and aims of a phenomenological ontology—that discipline which he claims as his own in the sub-title of *Being and Nothingness*. Unlike Heidegger, who attempted and failed to come to the question of Being, Sartre remains firmly locked within the anthropological purview of man. Sartre's solution to this problem, namely, ". . . the For-itself and the In-itself are reunited by a synthetic connection which is nothing other than the For-itself itself" (*BN*, p. 785), is wholly unsatisfactory in our view because it reduces ontology to anthropology.

3. Sartre's analysis of the structures of *réalité humaine* are certainly more creative than either Kierkegaard's or Heidegger's. But we may ask whether they describe the affective manifoldness of human existence. It seems to us that Sartre's depictions of this affective side, discussed under such rubrics as love, language, masochism, indifference, desire, hate, and sadism, are clearly consistent with the general argument of *Being and Nothingness*, but they are not, at least in our opinion, apodictically certain truths about the full nature of being human. On the contrary, such descriptions appear to follow as constructs, antimonies of the negativity inherent in Sartre's interpretation of the pre-reflective *cogito*: therefore, they are essentially ambiguous. But more importantly, being locked into a consciousness in which there is no content means ultimately to be cut off *solus-ipse* from all other beings (human or otherwise) completely. Sartre's attempt to circumvent solipsism, namely, ". . . the Other is the indispensible mediator between myself and me. I am ashamed of myself *as I appear* to the Other" (*BN*, p. 302), is undermined by the phenomenon of Angst which discloses that we are each the source of our own values. How then *can* the other shame us?

It is not necessary to belabor our point. There are so many problems and promises in Sartre's work that several excellent books have been devoted to these problems alone. The few issues we raise here seem to us, however, to be grounded in the ultimate presupposition of an *Ontos* without *Telos*, or to use the language of theology, an order of Being without God. This vast display of what absolute freedom means in the face of a Godless universe is necessary, of course, to Sartre's general argument in *Being and Nothingness*, namely, ". . . to explain man's predicament in human terms without postulating an existent God to guarantee anything."/50/ The God Sartre denies, it must be remembered, is the theistic God of Christian and Hebraic thought as interpreted by the Scholastics: "a specific all powerful, absolute, existing Creator."/51/ It is clear, one might even say "apodictically certain," that God's existence can no more be *disproven* logically than proven logically. For this reason the existence of God has rightly been called a matter of faith. But Sartre has a faith as well: that there exists no omniscient, omnipotent, omnibenevolent creator and sustainer of Being, and the argument of *Being and Nothingness* works out the consequences of Sartre's faith. Such is the apex of Sartre's atheistic "existentialist" position. Hence Angst in Sartre differs from Angst in Kierkegaard for obvious reasons; but Sartre's Angst concept also differs from Heidegger's and for less obvious reasons. As we shall see, Heidegger has a hidden agenda behind his published works: an onto-theological status for Being. Such an idea could never occur to Sartre, and thus his atheistic Angst differs at root from Heidegger's onto-theological Angst.

If at bottom Sartre's depiction of primordial Angst is unique because of his atheistic "faith," we shall require a theistic approach to primordial Angst apart from which Sartre's may be contrasted. Such is the task to which we now turn—namely, how primordial Angst is revealed in Paul Tillich's two works, *Systematic Theology* and *The Courage to Be*.

Part 2. Tillich on Angst

We choose Tillich as a contrast to Sartre on three grounds: (1) because of his lifelong enchantment with Angst in the face of the abyss, the unifying metaphor of this essay; (2) because of his clear working-out of the theological implications of Heidegger's conception of Angst and Nothingness; and (3) because of the clarity, laser-like precision, and power with which Tillich describes and interprets primordial Angst, or what he calls "ontological anxiety" in both his *Systematic Theology* and *The Courage to Be*. Clearly the role of Angst is as vital to these two works as it is to Kierkegaard's *Begrebnet Angst*, Heidegger's *Sein und Zeit*, and Sartre's *L'Etre et le neant*. Thus Tillich's descriptions and interpretations of the phenomenon are essential to our efforts here.

The fact that Tillich uses the term "ontological anxiety" in the place of primordial Angst need trouble us no more than Sartre's usage of *angoisse* for Angst. Yet, we must not mistake "ontological anxiety," as Tillich views it, for what we ourselves mean by "primordial Angst." As we shall see, the concepts are related but not identical, a fact observed by Rollo May in his brief biographical tribute to Tillich entitled *Paulus*. May notes Tillich had a profound sensitivity to Angst in the face of the abyss, and, moreover, that Tillich believed this Angst was close to what Heidegger meant by primordial Angst. Says May:

> Living with Angst, as it is called in his [Tillich's] mother tongue, is part of the price of loving the abyss. Tillich believed that anxiety was the subjective side of the tension between being and nonbeing in a sense similar to Heidegger's and it was therefore inevitable. The Angst could attach itself to almost anything in his existence./52/

And, in another passage May says:

> Paulus believed that every encounter with a new person is anxiety-creating. Anxiety is present in every authentic encounter, that is, one in which people let themselves genuinely meet. This anxiety is the dread of freedom; in Kierkegaard's words it is the "dizziness of freedom."/53/

From the standpoint of hermeneutic phenomenology, no matter how interesting and provocative these passages are, they may serve only as a prelude to an interpretation of "ontological anxiety" in Tillich's sense of that word. Again, we must not be guilty of blindly accepting May's authority to validate any interpretation of Angst in Tillich's thought, all the more so since that is precisely what we *accused* May of in the introduction./54/ Thus, a hermeneutical inquiry into the role of Angst in Tillich's thought is necessary. We do believe, however, that as with all displays of the Angst phenomenon, Tillich's "ontological anxiety" is likewise a facet of what we have called primordial Angst. Thus we hope to show here how primordial Angst is revealed as what makes possible anything like the "ontological anxiety" discussed by Tillich.

We turn now to a depiction of the horizons of influence encompassing Tillich's mature existential philosophy. While space limitations preclude any *full* treatment of the background, a treatment which incidentally has been undertaken with partial success by David Hopper in his work *Tillich: A Theological Portrait*,/55/ we hope to explore several major influences on Tillich's thought that made possible the disclosure of Angst in his thought. It is to this precise task that we now turn.

The Horizons of Influence

Any attempt to grasp the overall horizons of influence upon Tillich's mature thought must begin with at least a passing reference to his intellectual awakening. Tillich recollects:

> In Koenigsberg as well as in Berlin, I was a pupil in a "humanistic *Gymnasium*." . . . A humanistic *Gymnasium* has as its central subjects Greek and Latin. My love of the Greek language was a vehicle for my love of Greek culture and especially the early Greek philosophers. One of my most enthusiastically prepared and best received courses had as its subject matter the pre-Socratic philosophy./56/

Another early influence was Tillich's father, Johannes Tillich, a minister and administrator in the Prussian Territorial Church during his son's childhood. He established within the young Tillich a burning desire for personal and intellectual autonomy. This was accomplished by intense philosophical discussions between father and son, discussions that Tillich judged to be ". . . the most happy instances of a positive relation to my father."/57/ Tillich's early and deep attachment to philosophy during his years at Halle as a theology student was one major outcome of this father-son relationship. But even before these years, specifically with regard to the years between 1900 and 1905, Tillich tells us:

> Long before my matriculation as a student of theology, I studied philosophy privately. When I entered the University, I had a good knowledge of the history of philosophy, and a basic acquaintance with Kant and Fichte. Schleiermacher, Hegel, and Schelling followed, and Schelling

> became the special subject of my study. Both my doctoral dissertation and my thesis for the Licentiate in Theology dealt with Schelling's philosophy of religion./58/

Thus we find the first philosophers to strongly influence Tillich were precisely those to whom he remained committed throughout his life, especially Schelling./59/ Tillich's doctoral dissertation at the University of Breslau was entitled, *Die religionsgeschichtliche Konstruktion in Schellings positiver Philosophie, ihre Voraussetzungen und Prinzipien*/60/ [*The Construction of the History of Religion in Schellings's Positive Philosophy*]. One year after its successful defense in 1911, Tillich wrote a second dissertation for the Licentiate in Theology degree at the University of Halle entitled, *Mystik und Schuldbewusstsein in Schellings philosophischer Entwicklung*/61/ [*Mysticism and Guilt-consciousness in Schelling's Philosophical Development*]. This second dissertation appeared to have greater philosophical significance for Tillich: he included it in his *Gesammelte Werke* but excluded his doctoral dissertation./62/ What then is the gist of each work? The first seems to be a straight-forward account of Schelling's positive philosophy as well as an exegesis of themes relating to the development of Schelling's history of religion./63/ The second work, on the other hand, appears to be far more sophisticated, attempting with some flair to present Schelling's philosophical growth in the light of the antimony between guilt-consciousness as a function of *practical reason* on the one hand, and mysticism as the realization of *theoretical reason* on the other. This basic cleavage drives other antimonies/64/ becoming ultimately the radical distinction between philsophy and religion. Tillich argues that only if the *will*, as the power of self-contradiction (formal freedom), is made the ultimate metaphysical principle, can these antimonies be overcome. Thus, ". . . Schelling's philosophical development is represented as a dialectical advance toward the realization of this conclusion."/65/

What specifically did Tillich appropriate from Schelling's later thought? First, Tillich believed he had discovered in Schelling's will principle the roots of a synthesis between philosophy and religion. Secondly, Tillich saw in Schelling's mature period a Christian philosophy of existence that stood in direct contrast to Hegel's humanistic philosophy of essences./66/ Third, and perhaps most important to the present work, Tillich saw in the later Schelling a decisive break with Hegel's Idealism that made possible Søren Kierkegaard's discovery of existential philosophy's roots. Regarding this last point, we learn that "Schelling, according to Tillich, had seen the chasm which looms before man but had quickly averted his eyes from the terrifying sight. Kierkegaard looked at it unflinchingly."/67/ Again, the abyss metaphor.

With respect to other significant influences, at this time Tillich became acquainted with phenomenology through reading Husserl's *Logische Untersuchungen*. Husserl deeply impressed the young Tillich, who confessed later that Husserl's thought became a most satisfying confirmation of his knowledge of Kant and Fichte./68/ Be that as it may, Tillich could not attach himself to Husserl's phenomenology because it "lacked the dynamism"/69/ necessary to his own manner of thinking. Tillich enthusiastically approved, however, of Husserl's refutation of positivism and psychologism.

Regarding his teachers at Halle, clearly the most influential was the theologian Martin Kahler. Kahler showed Tillich ". . . the all-embracing character of the Pauline-Lutheran idea of justification."/70/ This insight into justification through faith led Tillich to a keen appreciation of man's finitude, his guilt and despair, and his ultimate estrangement from God as the source of his Being. Kahler likewise led Tillich to see the meontic *Ungrund* of the abyss within the God-man relationship, wherein man as sinner is justified before God's judgment. Moreover this insight, Tillich tells us, led him to accept the existential ideas revealed in the respective philosophies of Søren Kierkegaard and later Martin Heidegger./71/ Tillich tells us that because of Kahler, ". . . it was easy for me . . . to accept the analysis of human existence given by Kierkegaard and Heidegger."/72/ Clearly, however, Kierkegaard's influence on Tillich predated that of Heidegger by at least fourteen years, since Kierkegaard is mentioned in Tillich's second dissertation, written in 1913. We know that Tillich was not aware of Heidegger's work until 1925./73/

We see then that during an early period of his theological and philosophical development Tillich became acquainted with Kierkegaard's rediscovered thought. Tillich says that while several important thinkers shaped his theological perspective during his student years, outstanding among them was ". . . our discovery of Kierkegaard and the shattering impact of his dialectical psychology. It was a prelude to what happened in the 1920s when Kierkegaard became the saint of the theologians as well as the philosophers."/74/ This profound historical record goes a long way toward explaining the general intellectual climate of opinion under which both Heidegger and Tillich (who were roughly contemporaries) wrote regarding Angst. As we have seen Heidegger began *Sein und Zeit* prior to 1925 and Tillich tells us that he began Volume One of the *Systematic Theology*, wherein his first treatment of Angst is couched, in 1925./75/ Clearly the spirit of Kierkegaard's *Begrebnet Angst* filled the air of post-World War I Germany.

During his brief stay at Marburg, one which lasted only three semesters, Tillich very likely saw Heidegger (who was in residence there as an *ordinarius* professor). According to Tillich's wife Hannah, however, these two intellectual giants never met formally./76/ As she saw it, there existed a marked tension between her husband and Heidegger. Mrs. Tillich states that her husband's main struggle at Marburg

> . . . was with the philosophy of Heidegger. He met some of Heidegger's doctoral students, and endless debates followed. Oddly enough, the Ordinarius Professor Heidegger and the private dozen Adjunct Professor Tillich never met during our stay in Marburg. The gossip about what Heidegger had said in a lecture about Paulus/77/ would be carried by Heidegger's faithful underlings. Paulus would answer in his own lecture, and that would be forwarded again. In time, it resulted in Paulus's growing influence—at the end of this second term, he had a full lecture hall—but when he left for Berlin, having survived the term, he fainted in the corridor of our third-class compartment. The experience in Marburg had been grim./78/

If we get beyond the pettiness of this lack of encounter, we see the dark undercurrent of distrust that developed between Tillich and Heidegger. Yet Tillich himself was to say much later of this period:

> In Marburg, in 1925, I began my work on my *Systematic Theology*, the first volume of which appeared in 1951. At the same time that Heidegger was in Marburg as a professor of philosophy, influencing some of the best students, existentialism in its twentieth century form crossed my path. It took years before I became fully aware of the impact of this encounter on my own thinking. I resisted, I

tried to learn, I accepted the new way of thinking more than the answers it gave./79/

In *On the Boundary* Tillich provides the following account of Heidegger's thought. First, he sees Heidegger's philosophy as a doctrine of man, one which although unintentional is at once a doctrine of human freedom and human finitude. Second, from Tillich's perspective Heidegger's thought in *Being and Time* ". . . is so closely related with the Christian interpretation of human existence that one is forced to speak of a 'theonomous philosophy,' in spite of Heidegger's emphatic atheism."/80/ As we shall see, this is important in terms of understanding the special "theism" of Tillich.

Thus Tillich was able to summarize his relationship to his own later existential philosophy/theology in terms of the combined influence of Schelling, Kierkegaard, Husserl, and Heidegger. In an important passage, Tillich talks about his introduction to modern existential thought. He says:

> When existential philsophy was introduced to Germany, I came to a new understanding of the relationship between theology and philsophy. Heidegger's lectures at Marburg, the publication of *Sein und Zeit* (*Being and Time*), and also his interpretation of Kant were significant in this connection. Both to the followers and to the opponents of existential philosophy, Heidegger's work is more important than anything since Husserl's *Logische Untersuchen* (*Studies in Logic*)./81/

Coming from the leading spokesman for existential theistic thought, this is a high complement to one who many saw as a flagrant atheist. But as far as we are aware, Heidegger never acknowledged Tillich as a thinker, much less returned the complement. In any event, Tillich continues the above passage with an account of how he came to accept existential thought:

> Three factors prepared the ground for my acceptance of existential philosophy. The first was my close knowledge of Schelling's final period, in which he outlined his philosophy of existence in response to Hegel's philosophy of essence. The second was my knowledge, however limited, of Kierkegaard, the real founder of existential philosophy. The final factor was my enthusiasm for Nietzsche's "philosophy of life." These three elements are also present in Heidegger. Their fusion into a kind

> of mysticism tinged with Augustinanism accounts for the fascination of Heidegger's philosophy./82/

We learn from this passage the deep influence Nietzsche had on Tillich's thought. While Tillich did not read Nietzsche until 1916, Nietzsche's influence on Tillich was especially important regarding Tillich's fondness of the abyss metaphor. Tillich says, "Nietzschean vitalism expresses the experience of the abyss more clearly than Neo-Kantianism, value-philosophy, or phenomenology."/83/ We shall return to this central concept of the abyss in Tillich's thought in the following sections, but we should recognize here that Nietzsche made a profound contribution to Tillich's understanding and interpretation of the abyss metaphor.

Likewise during his time at Marburg, Tillich frequently met with Rudolph Otto. Tillich had been influenced early on by Otto's idea of the *mysterium tremendum*, and the two often met socially over coffee and cakes. Otto, who in the vision of Hannah Tillich was ". . . the sky-blue-eyed, white-haired, slim and youthful-looking man who talked about India, about beauty, about art,"/84/ touched Tillich at the deepest levels of his theological existence. Tillich says that Otto's work, *Das Heilige* [*The Idea of the Holy*], determined his entire method in the philosophy of religion./85/ Specifically, from Otto's work Tillich came to see that he must *start* with the experience of the holy and *move* to the idea of God. But, says Tillich, equally important to this theological development was Otto's comprehension of the mystical, sacramental, and aesthetic implications derivable from the idea of the holy. More specifically, Tillich learned form Otto that the ". . . ethical and logical elements of religion were derived from the *experience* of the presence of the divine, and not conversely."/86/ It was, therefore, a phenomenology of the Holy that intrigued Tillich.

Certainly the phenomenology of the holy represents an important connecting link between Otto and Tillich. This link demonstrates that both accepted a religious phenomenology, a genuine phenomenology derived independently from Husserl and one grounded in Kant's schematization of the categories as reinterpreted by Jakob Friedrich Fries (1773-1843). It is important to mention Fries as one who must have been an influence on Tillich: for Tillich's friendship with Otto makes it certain that the former was aware of Otto's work of 1909 entitled *Kantische-Fries'sche Religions philosophie* (which in 1932 was translated as *The Philosophy of Religion based upon Kant and Fries*, but which in our view has been seriously ignored).

Otto's influence clearly left its mark on Tillich. Years later, in his work entitled *My Search for Absolutes*, Tillich was to recall with obvious warmth:

> When I see . . . [the] phrase "the Idea of the holy," I remember wonderful hours in Marburg, Germany, in the mid-twenties, when Rudolf Otto and I walked together through the hills and woods and talked about the problems of Christianity and the Asiatic religions. . . . The first thing he said in his analysis of the meaning of the term "holy" was that "the holy" is a "mystery" and means the Absolute itself, the ground of all the absolutes we have discovered in the different realms of man's encounter with reality. . . . In experiencing this mystery man is driven to ask the question: "Why is there something and not nothing?"/87/

Here emerges again the familiar echo of Heidegger's question of the essence of ground, but from a completely different source, namely, Otto's *Das Heilige* and the "aweful mystery" of the Holy. Tillich goes on to describe the phenomenological approach to Otto's *mysterium tremendum*:

> Otto expresses the relation of our mind to the Ultimate and its mystery in two terms: "*tremendum*"—that which produces trembling, fear and awe; and "*fascinosum*"—that which produces fascination, attraction and desire. . . . The dread of missing one's fulfillment—this is awe. The desire to reach one's fulfillment—this is attraction./88/

To this description we must add one other that decisively connects Tillich's conception of the holy with Otto's. In *Systematic Theology*, Tillich says of Otto's notion of the holy:

> When he [Otto] describes the mystery of the holy as *tremendum* and *fascinosum*, he expresses the experience of "the ultimate" in the double sense of that which is the abyss and that which is the ground of man's being. This is not directly asserted in Otto's merely phenomenological analysis, which, by the way, never should be called "psychological." However, it is implicit in his analysis, and it should be made explicit beyond Otto's own intentions./89/

From this important passage we discover that Otto's notion of the *tremendum*, the "numinous dread or awe,"/90/ struck Tillich as the key to the abyss and the ultimate. This represents an important insight for Tillich, one intimately connected with how primordial Angst, as ontological anxiety, is revealed in Tillich's thought.

In summary, we see that from the later Schelling, Tillich secured the possibility of a higher synthesis into which both philosophy and theology could be merged. From Kahler he gained the central insights necessary to embrace existential thinking. From Kierkegaard, Tillich appropriated an essential grasp of existential theology; from Husserl he derived an appreciation for the method of phenomenology and the refutation of both positivism and psychologism. From Heidegger he derived the radical insight into man as finitude as well as an appreciation for the canons for modern existential thinking. From Nietzsche he secured a deep affinity for the abyss metaphor as it applies to human finitude and Philosophy of Life (*Lebensphilosophie*). And finally, from Otto he acquired the understanding of the *abyss* as the *tremendum experience* of the Absolute,/91/ although Otto may not have seen it as such.

With these influences to guide him, Tillich was able to direct his energies towards concretely working-out the phenomenon of primordial Angst. Specifically for Tillich, Angst is: (1) man's awareness of his ownmost finitude;/92/ and (2) the essential condition of making possible man's ontological courage, or more precisely, ". . . existential awareness of non-being"/93/ that man must face courageously. Let us examine these descriptions now in some detail.

Some Relevant Elements of Systematic Theology

As we have seen in the preceding section, Tillich began his *Systematic Theology* in 1925 at Marburg. The finished work of Volume I, however, was not published until 1951, over a quarter of a century later. Beyond dispute Tillich's depictions of Nothingness, nonbeing, and even Angst itself postdate Heidegger's analyses and interpretations in *Sein und Zeit*. This *must* be true because Tillich not only refers to Heidegger's interpretations, but *uses* them to explain and support his own views. This fact is important. It displays concretely Tillich's debt to Heidegger regarding theonomous philosophy, the existential underpinnings of Christian thought which take their point of departure from the analysis of man's radical finitude. Let us see how this is so, first with regard to the *Systematic Theology* and then in regard to *The Courage to Be*.

The work *Systematic Theology* was Tillich's life-long project. Its specific purpose, says Tillich, was

> . . . to present the method and structure of a theological system written from an apologetic point of view and carried through in a continuous correlation with philosophy. The subject of all sections of this system is the method of correlation and its systematic consequences illustrated in a discussion of the main theological problems. (*ST*, "Preface," p. xi)

But Tillich reserved the general purpose of the work for the last powerful line of his preface. There he says, "A help in answering questions: this is exactly the purpose of this theological system" (*ST*, "Preface," p. xii).

Since *Systematic Theology* contains over 900 coherently related pages, we cannot possibly explore must less display its full range here. At bottom, however, Tillich expresses a fundamental distinction between philosophy and theology which necessitates a *systematic theology* that can answer the existential questions raised by philosophy. This he calls "theonomous philosophy." For at root, Tillich argues, philosophy and theology differ radically. This position of course represents a major shift from his 1911 Halle dissertation that sought to unify these two disciplines./94/ Thus in the *Systematic Theology* Tillich defines philosophy as "that cognitive approach to reality in which reality as such is the object" (*ST*, I:18). This view implies that for Tillich, philosophy must be specifically identified with ontology. Ontology, then, Tillich defines as ". . . an analysis of those structures of being which we encounter in every meeting with reality" (*ST*, I:20). Accordingly, ontological analyses are *confirmed* in philosophy via experimental verification, by which Tillich means ". . . the way of an intelligent recognition of the basic ontological structures within the encountered reality, including the process of encountering itself."/95/

Theology is likewise ontologically grounded, says Tillich, but seeks to display ". . . the *meaning* of being for us" (*ST*, I:20). Thus the theologian has a markedly different viewpoint towards his discipline. Specifically the theologian

> . . . is involved—with the whole of his existence, with his finitude and his anxiety, with his self-contradictions and his despair, with the healing forces in him and in his social situation. Every theological statement derives its seriousness from these elements of existence. The theologian, in short, is determined by his faith. (*ST*, I:23)

From this primary difference other important differences follow, namely: (1) while the philosopher attempts to discover the structure of reality as a whole, the theologian attempts to discover the meaning of a historical event; namely, the coming of Christ, in which his "ultimate concern"/96/ is manifest; (2) while the philosopher deals wholly with the categories of Being, such as physical and psychological causality, or biological and historical time, for example, the theologian ". . . related the same categories and concepts to the quest for a 'new being'"/97/ (*ST*, I:24).

Yet, there are likewise confluences of philosophy and theology: (1) The philosopher *and* the theologian both exist as human beings in a world. Hence the former cannot ". . . jump over the concreteness of his existence and its implicit theology" (*ST*, I:24) any more (or less) than can the theologian. (2) Moreover, and perhaps even more importantly, the philosopher, like the theologian ". . . exists in the power of an ultimate concern, whether or not he admits it to himself and to others" (*ST*, I:24). In any case, says Tillich, the philosopher *as* philosopher *should* admit it, because ". . . without an ultimate concern his philosophy would be lacking in passion, seriousness, and creativity" (*ST*, I:25). At rock bottom, says Tillich, "every creative philosopher is a hidden theologian . . ." (*ST*, I:25). Why? Because the philosopher both as thinker and as human being must have the existential passions or what Tillich calls "ultimate concerns" (*ST*, I:27) and the rational power to carry out his philosophical commitment.

Despite these confluences Tillich concludes that a common basis is lacking between philosophy and theology. Philosophy, on the one hand, looks for the hidden structures of Being. Theology, on the other hand, deals with the existential passions or man's *ultimate concerns* in the light of such structures. To understand, therefore, the place of philosophical inquiry, and especially Angst's place within Tillich's system, we must first discern how

Tillich conceives of the role of reason in general. Tillich's discussion of reason in man leads us directly to his analysis of our root metaphor, Angst and the abyss.

For Tillich "reason" has three distinct functions. First, "reason" has the sense of "technical reason," by which is meant the calculative, self-consciously logical, scientific, and analytical mode of thinking. Secondly, there is "ontological reason," by which Tillich means ". . . the structure of the mind which enables the mind to grasp and transform reality" (*ST*, I:22–23). Finally, there is "ecstatic reason" which is grounded, Tillich tells us, in Christian revelation. As "ecstatic" reason it is "self-transcending" power that enables us to grasp revelations as manifestations of the meontic mystery of the ground of Being. The antimonies and outright contradictions to which both technical and ontological reason are subject *can be overcome only* by ecstatic reason. It is therefore the source of truth, that is, existential truth with regard to man's ultimate concerns.

We must keep in mind that ecstatic reason is grounded in the ultimate meontic mystery of Being itself. Thus when ecstatic reason discloses the meaning of our ultimate concerns, it springs from the abysmal depths of the mystery of Being, the meontic *Urgrund*. When this occurs, according to Tillich, man undergoes a profound "ontological shock." We believe "ontological shock" may be another form of a primordial Angst in the face of the abyss. Specifically, in ecstasy, Tillich tells us, the rational mind ". . . transcends the basic condition of finite relationality, the subject-object structure" (*ST*, I:112). "Ontological shock" occurs as the negative side of our ecstatic penetration into the meontic mystery of Being. Tillich says:

> The threat of nonbeing, grasping the mind, produces the "ontological shock" in which the negative side of the mystery of being—its abysmal element—experienced. "Shock" points to a state of mind in which the mind is thrown out of normal balance, shaken in its structure. Reason reaches its boundary line, is thrown back upon itself, and then is driven again to its extreme situation. This experience of ontological shock is expressed in the cognitive function by the basic philosophical question, the question of being and non-being. (*ST*, I:113)

In ecstatic reason the ontological shock is both preserved and overcome. It is preserved in the annihilating power of the holy (*mysterium tremendum*) yet preserved in the elevating power of the divine presence (*mysterium fascinosum*). Accordingly, Tillich tells us that "Ecstasy unites the experience of the abyss to which reason in all its functions is driven with the experience of the ground in which reason is grasped by the mystery of its own depth and of the depth of being generally" (*ST*, I:113). Thus we see that ecstatic reason is itself *grounded in the meontic depths of the abyss* as the source of our power to solve the antimonies of ontological reason on the one hand and technical reason on the other. As such, the limits placed upon ecstatic reason by human finitude show themselves through the phenomenon of profound Angst experienced as the "ontological shock," the negative side of the ecstatic penetration into the mysterious abyss of Being.

Part two of *Systematic Theology* deals with "Being and the Question of God." Following his established method of "continuous correlation," Tillich begins with the question of Being as a philosophical question. But the very raising of this question is connected with the profound Angst that first arises in "ontological" or "metaphysical" shock. Tillich states: "the ontological question, the question of being itself, arises in something like a 'metaphysical shock'—*the shock of possible non-being*" (*ST*, I:163) (italics added). And, moreover, when this question arises, ". . . everything disappears in the abyss of possible non-being; even a god would disappear if he were not being-itself" (*ST*, I:164).

From Tillich's perspective, then, ontology is possible only because some concepts are *less* universal than Being itself, but still *more* universal than a lower level set of concepts relating to ontic, individual beings. These intermediary concepts, says Tillich, have been called "categories," or "principles," or even "ultimate notions" (*ST*, I:164); but, says Tillich, no agreement has been reached throughout the history of philosophy as to precisely which of these concepts *must* be included in any full-dress ontology ". . . although certain concepts reappear in almost every ontology" (*ST*, I:164).

To aid in understanding the fundamental aegis of ontology, Tillich suggests that there are at least four distinct *levels* of ontological concepts which can be meaningfully distinguished. These are: (1) the basic ontological structure implied in the question of being; (2) the precise elements constituting that structure; (3) the characteristics of being that make possible human existence; and finally, (4) the explicit categories of Being and knowing. We will return to these later in the light of Tillich's explicit discussion of Angst.

In a subsection entitled "A. The Basic Ontological Structure—Self and World" (*ST*, I:168), Tillich observes: "Every being participates in the structure of being, but man alone is immediately aware of this structure" (*ST*, I:168). But why man and man alone? Because, says Tillich, man is the only being *who can question Being*. In this pronouncement Tillich agrees with Heidegger's definition of Dasein as well as Sartre's conception of *réalité humaine*: man's being is the only kind of being that can question. Says Tillich, "man occupies a pre-eminent position in ontology . . . as that being who asks the ontological question and in whose self-awareness the ontological answer can be found" (*ST*, I:168). And speaking of Dasein specifically, Tillich notes that Dasein is *given* to man within himself—man can answer the ontological question only because he experiences being in an immediate and direct manner (*ST*, I:169–70). Man, therefore, is that being who is aware of Being's structures. This, according to Tillich, makes cognition possible to begin with.

But for Tillich, man is not only the "being-there" of Dasein. Rather, man possesses a Self—a self which "has" experiences, and is bound in a relationship of concern to his ontic world. Thus the Self is, says Tillich, ". . . more embracing than the term 'ego.' It includes the subconscious and the unconscious 'basis' of the self-conscious ego as well as self-consciousness (*cogitatio*) in the Cartesian sense" (*ST*, I:169). In adopting this perspective, Tillich departs from the view of Heidegger's Dasein in a most radical and extremely far reaching way by postulating the existence of human subconsciousness, and even an unconscious element within the Self./98/

Man in Tillich's view is ". . . a fully developed and completely centered Self. He 'possesses' himself in the form of self-consciousness. He has an ego-self" (*ST*, I:170). Thus, man can transcend every possible environment as an ego-self, because he *has* the concept of world and, at the same time, is *in* the world: the structural unity of experience's manifoldness. Nonetheless, man is ultimately finite. "World-consciousness," says Tillich, "is possible only on the basis of a fully developed self-consciousness" (*ST*, I:171). We must, therefore, examine this essential finitude in order to understand how man *has* any world to begin with. Such an analysis leads us finally to primordial Angst in Tillich's thought.

Primordial Angst as "Anxiety" in Systematic Theology

Tillich repeatedly insists that the question of Being ". . . is produced by the 'shock of nonbeing'" (*ST*, I:186). Being itself must remain an ultimate and meontic mystery. But man has access to Being by his ability to "envisage nothingness" (*ST*, I:186). The history of man's encounter with Nothingness culminates in modern existential thinking—the culmination of a long and arduous task of unfolding the meaning of Being based on man's ability to "envisage Nothingness." Tillich says:

> Recent existentialism has "encountered Nothingness" (Kuhn) in a profound and radical way. Somehow it has replaced being-itself by non-being, giving to nonbeing a positivity and a power which contradict the immediate meaning of the word. Heidegger's "annihilating Nothingness" describes man's situation of being threatened by nonbeing in an ultimately inescapable way, that is, by death. The anticipation of Nothingness at death gives human existence its existential character. Sartre includes in nonbeing not only the threat of Nothingness but also the threat of meaninglessness (i.e., the destruction of the structure of being). In existentialism there is no way of conquering this threat. (*ST*, I:189)

Such is Tillich's introduction to the threat of Nothingness in *Systematic Theology*. Here we see that nonbeing is experienced in man as the "not yet" as well as the "no more" of Being. Being, which in man is limited by nonbeing, is experienced as human finitude (*ST*, I:189).

According to Tillich, finitude is encountered when I become aware of non-being's potential threat. This threat is revealed, says Tillich, in "ontological anxiety." Such Angst has an *ontological* quality; it cannot be derived from something more primordial. As such, Angst or ontological anxiety is as radical as my finitude; for Angst is dependent on the threat of my nonbeing as "inward" apprehension of my "outward" finitude. Like Kierkegaard, Heidegger, and Sartre before him, Tillich agrees that there can be no direct object of Angst: only in fear is there such an object. "A danger, a pain, an enemy, may be feared," says

Tillich, "but fear can be conquered by action" (*ST*, I:191). Angst, on the other hand, can never be conquered; for no finite being can ever conquer the inward apprehension of its own finitude. So Angst ". . . is always present, although often it is latent. Therefore, it can become manifest at any and every moment, even in situations where nothing is to be feared" (*ST*, I:191). To this last passage Tillich adds an important footnote regarding psychological versus ontological Angst. Tillich states: "psychotherapy cannot remove ontological anxiety, because it cannot change the structure of finitude" (*ST*, I:191, footnote 7). Psychotherapy *can*, however, put anxiety in its proper place (by removing its compulsive manifestations and thereby reducing the frequency and intensity of ontic fears); but it can *never* remove ontological Angst.

The distinction between Angst as ontological anxiety and "compulsory forms of anxiety" cannot be overemphasized or undervalued. To decisively sharpen this distinction, we draw attention to the following crucial passage in *Systematic Theology*; for here the ontological dimension of primordial Angst is contrasted once and for all with pathological anxiety. Pathological anxiety, says Tillich, is simply another manifestation of *fear*, an ontic phenomenon. Tillich tells us why:

> The recovery of the meaning of anxiety through the combined endeavors of existential philosophy, depth psychology, neurology, and the arts is one of the achievements of the twentieth century. It has become clear this fear as related to a definite object and anxiety as awareness of finitude are two radically different concepts. Anxiety is ontological; fear psychological. (*ST*, I:191)

Here Tillich adds an important footnote concerning the terms "Angst" and "anxiety" as distinguished from "dread."/99/ Specifically, the English word "anxiety" has taken on the connotation of the German word Angst only during what was at that time the previous decade: *the decade of World War II*. In the footnote Tillich states that both Angst and anxiety are derived from the Latin "augustiae," which, as we previously saw in our Introduction, means "narrows." Tillich concludes: "anxiety is experienced in the narrows of threatening Nothingness. Therefore, anxiety should not be replaced by the word 'dread,' which points to sudden reaction to a danger but not to the ontological situation of facing nonbeing" (*ST*, I:192, footnote 8). This is what we ourselves called strictural Angst in Chapter I.

Thus Angst for Tillich is an ontological phenomenon expressing man's finitude as ". . . the self-awareness of the finite self as finite" (*ST*, I:192). This means that Tillich's conception of Angst, like Sartre's conception of *angoisse*, is a reflective apprehension of man's finitude in the face of Nothingness or nonbeing, a point-of-view that should provide fertile ground for contrasting Sartre and Tillich on Angst, a topic reserved for our final chapter.

Angst and the Categorical Forms of Finitude

Tillich's full display of primordial Angst's meaning can take place only from within the categories "internal to man's being; namely, time, space, causality, and substance." Specifically, such categories are in Tillich's view ". . . the forms in which the mind grasps and shapes reality" (*ST*, I:192). But these categories are not simply logical forms which determine *discourse* about reality. Rather they determine our knowledge of reality, because ". . . the mind cannot experience reality except through the categorical forms" (*ST*, I:192). Tillich states:

> The categories reveal their ontological character through their double relation to being and nonbeing. They express being, but at the same time they express the nonbeing to which everything that is, is subject. The categories are forms of finitude; as such they unite an affirmative and negative element. . . . Each category expresses not only a union of being and nonbeing but also a union of anxiety and courage. (*ST*, I:192–93)

For Tillich the central category of finitude is time, which throughout the history of Western philosophy has displayed both the positive and negative elements of being and non-being. First, with regard to time's *negative* elements, philosophy has both shown the transitoriness of everything temporal and the impossibility of fixing the present "now" within the temporal flux. With regard to time's positive elements, philosophy has seen the creative, direct, and irreversible character of time. But at rock bottom, philosophical ontology can only *observe* the balance between these antipodal views; it can never decide the meaning of time in itself (*ST*, I:193).

Experienced "inwardly" (i.e., within immediate self-awareness) time is the category that ". . . unites the anxiety of the transitoriness with the courage of a self-affirming present" (*ST*, I:193). Tillich points out that my melancholy awareness of my necessity to die, my internal ontological anxiety about *having* to die, is what reveals to me the external ontological character of time. "In the anxiety of having to die nonbeing is experienced from 'the inside.' This anxiety is potentially present in every moment. It permeates the whole of man's being; it shapes soul and body and determines spiritual life; it belongs to the created character of being quite apart from estrangement and sin" (*ST*, I:193-94).

Tillich's theological answer to the Angst associated with transitoriness is a *courage* to affirm temporality. Without such courage I would succumb to the annihilating character of time. Man needs such courage because in order to maintain his being he must existentially conquer his deepest ontological anxiety by defending his present against the threat of no future.

Secondly, space is likewise a category which expresses the element of being and nonbeing. Put postively, "to be," says Tillich "means to have space" (*ST*, I:194). Obviously, the converse is likewise true: not to have space implies the threat of nonbeing. It is this latter formulation that gives rise to the phenomenon of "insecurity" as a manifestation of ontological Angst arising within the finite category of space. I *need* a physical space: ". . . the body, a piece of soil, a home, a city, a country, the world" (*ST*, I:194). Moreover, I need social space: ". . . a vocation, a sphere of influence, a group, a historical period, a place in remembrance and anticipation, a place within a structure of values and meanings" (*ST*, I:194). Such physical and social space, therefore, is an ontological necessity of man. The threat of its nonbeing is the "ultimate insecurity." But to be finite entails being radically insecure (*ST*, I:194). Thus I seek my own physical and social space with a passion, but my ontological anxiety shows through when I fully realize that because I am finite I must ultimately and literally lose my space in dying.

Again, courage is the theological answer to the Angst that threatens spacelessness. By affirming the present and the space given to me now, even if it is only for the moment, I am able to endure that which must ultimately be lost.

Third, with regard to causality, it too expresses being and

nonbeing. As for being, causality points to that which precedes something else as its source, brought to fruition through Being's power. Negatively, however, the cause-effect model displays the inability of any effect to be its own cause. This leads ultimately to the discovery that "things and events have no aseity" (*ST*, I:196). Angst of course shows itself in connection with this latter determination of causalilty, for man as a creature has only contingent being. He is, in Tillich's words ". . . the prey of nonbeing" (*ST*, I:196). Thus the ontological anxiety I discover in the the category of causality is revealed as Angst in the face of my lack of ontological necessity.

The theological solution is, again, courage. Here courage is in the form of an acceptance of contingency and derivedness; for "without this courage no life would be possible . . ." (*ST*, I:197). Specifically, if I have this courage I do not look beyond myself for my source of being; rather I find it within myself. Such courage, Tillich tells us, ignores the causal dependence of man's finite, contingent, and derived being.

Finally, regarding the category of substance or that which philosophy holds to underlie the flux of appearances, Angst is revealed when I realize that ultimately I must lose my substance each day, as an existential modification of my being. All this, of course, is intensified in the threat of ultimate nonbeing: the experience of my having to die. Thus, having to die, says Tillich, ". . . anticipates the complete loss of identity with one's self" (*ST*, I:198). This loss of identity is bound up with my threatened loss of the power to maintain myself as a coherent identity in a world of flux.

Again, courage is the theological solution to such ontological anxiety, specifically, the courage to make something substantial of what is ultimately accidental—for example, a creative work, an essay, a loving relationship, a career, a "lasting contribution," and so on. Existentially, *man creates values*. But again, theology is the answer here. Angst, as revealed in these four categories, can be overcome only by the fundamental and essential power that *lies behind the courage to face such Angst*. Says Tillich, "The question of God is the question of the possibility of this courage" (*ST*, I:198).

Angst and the Polar Elements of Being

Tillich concludes his discussion of Angst in *Systematic Theology* by showing the dynamic tension between Being's "polar elements" which transcend these categories. These polar elements are: (1) individualization and participation, (2) dynamics and form, and (3) freedom and destiny. The polar character of these elements is precisely what makes man vulnerable to the threat of non-being or Angst. Polarity, then, becomes *tension* under the impact of finitude. And by "tension" Tillich means, ". . . the tendency of elements within a unity to draw away from one another, to attempt to move in opposite directions" (*ST*, I:198).

My Angst in the dynamic tension of these polar elements is Angst in the face of

> . . . not being *what we essentially are*. It is anxiety about disintegrating and falling into nonbeing through existential disruption. It is anxiety about the breaking of the ontological tensions and the consequent destruction of the ontological structure. (*ST*, I:199) (Italics added.)

Personally, we can think of no circumstances more terrifying these last few words: "the consequent destruction of the ontological structure"; for such a *breaking* of the tensions implies a snap in the fine thread by which I am suspended over the bottomless abysses of meaninglessness, death, and moral condemnation.

In the first polar elements, namely, "finite individualization and finite participation," my being hangs suspended over the abyss, stretched between an ultimate loneliness on the one hand and a robot-like terror of becoming a collectivized being of lost individuality and subjectivity on the other. I shudder, hovering in Angst between these two poles, suspended over the abyss by a thin thread of Self that prevents my ultimate, unredeemable plunge into nonbeing. Discovering this dynamic tension, Tillich says, belongs essentially to depth-psychology and sociology, for the tension has been overlooked by philosophy. The exception, of course, was Pascal, who Tillich regards as the founder of the ontological underpinnings of the loneliness-belongingness polarity.

Secondly, Angst is revealed by the ontological polarities of "dynamics and form." Here again the abyss looms up before me. I am suspended on a thin thread of Self. On the one side, dynamics drives *toward* form ". . . in which being is actual and has the power of resisting nonbeing" (*ST*, I:199). Yet I am threatened because I may become *lost* in rigid forms. If I try to break these rigid forms, as indeed I must to attain balance, ". . . the result may be chaos, which is the loss of both dynamics and form" (*ST*, I:200). Again the abyss calls to me. I hover on a thin thread of Self between dynamics and form, dancing like a tightrope walker who has just lost his balancing bar. And the wind in the depths of the abyss is howling. Thus primordial Angst is revealed before ". . . the threat of a final form in which vitality will be lost . . . the threat of a chaotic formlessness in which both vitality and intentionality will be lost" (*ST*, I:200).

Finally, regarding the polarities of "freedom and destiny," the thin thread of Self extending over the abyss is especially tense. I am threatened on the one side by a loss of freedom in the face of finite necessities, and on the other side I am threatened by the loss of destiny, the countless contingencies that result from finite freedom. I can rush headlong into the notion of absolute freedom to save my finite freedom, and thereby lose my destiny. Or, I can fix my self firmly in my self-planned destiny and thereby concede my personal freedom. In such Angst freedom becomes arbitrariness: both freedom and destiny are lost. To lose our destiny, says Tillich, is to fall into the abyss of despair.

> To lose one's destiny is to lose the meaning of one's being. . . . Man's essential anxiety about the possible loss of his destiny has been transformed into an existential despair about destiny as such. Accordingly, freedom has been declared an absolute, separate from destiny (Sartre). But absolute freedom in a finite being becomes arbitrariness and falls under biological and psychological necessities. The loss of a meaningful destiny involves the loss of freedom also. (*ST*, I:201)

We will return to Tillich's refutation of Sartre's freedom doctrine again in our concluding chapter. But for now we conclude our discussion of Angst in *Systematic Theology* by citing a final passage to show why we see close correspondence between Tillich's conception of "ontological anxiety" and what we have called "primordial Angst." Specifically, in a footnote to his discussion of Angst and the dynamic polarity between freedom and destiny, Tillich observes:

> The material discussed in this chapter is by no means complete. Poetic, scientific, and religious psychology have made available an almost unmanageable amount of material concerning finitude not anxiety. The purpose of this analysis is to give *only an ontological description of the structures underlying all these facts* and to point to some outstanding confirmations of the analysis. (*ST*, I:201, footnote 9; italics added.)

This task, namely, providing an ontological description of Angst's primordial structure, is precisely our own. In the present essay we have tried to show how the effects of Angst are revealed in the primary works of the thinkers we have considered here, but at root we have tried to allow primordial Angst to speak for itself as the final ontological root that binds together such facets. We would argue, therefore, that since this root of Angst has become visible on an ontological level it must be valid at the ontic level, the level to which Tillich points in the passage cited above.

Primordial Angst as "Anxiety" in The Courage to Be

As with Heidegger and Sartre, Tillich's second major statement on Angst found expression through the lecture medium. In 1950 Tillich presented a series of lectures at Yale University which discussed the phenomena of courage and Angst. This series, and the book upon which they were based, was entitled *The Courage to Be*.

In this second statement the concepts of both Angst and courage are discussed in far greater detail than what was presented in *Systematic Theology*. In *The Courage to Be* both Angst and courage are examined more from the standpoint of *ethics* than from the standpoint of the ontology of theonomous philosophy; although to be sure both dimensions are clearly present. Tillich notes:

> The title of this book, *The Courage to Be*, unites both meanings of the concept of courage, the ethical and the ontological. Courage as a human act, as a matter of valuation, is an ethical concept. Courage as the universal and essential self-affirmation of one's being is an ontological concept. The courage to be *is the ethical act* in which man affirms his own being in spite of those elements of his existence which conflict with his essential self-affirmation. (*CB*, p. 3) (Italics added.)

Thus while the source of courage still requires absolute faith to carry ontological Angst within us, the "courage to be" is essentially an ethical act in which man affirms his own being "in-spite-of" the threats of non-being. This is where Angst comes in: Angst is precisely that ontological phenomenon which discloses these threats to man's being. So the subtle shift, from discussing Angst ontologically to this ethical approach, renders the discussion of Angst in *The Courage to Be* more "existentialistic" in the ethical, Sartrean sense of the term, than "phenomenological" in either Heidegger's or Otto's sense of *that* term. We are *not* claiming that Tillich has abandoned his position of *Systematic Theology*. Rather we are simply suggesting that the ethical, "existentialist" approach is more pronounced in *The Courage to Be* than in the approach of phenomenological ontology. But in our view this shift colors the entire interpretation of Angst at least one full tone—from a purely descriptive to a partially *prescriptive* or ethical discussion, making Angst a dark mirror for ethical courage.

What does this mean in terms of Angst's role in *The Courage to Be*? In *Systematic Theology* Tillich defined Angst as our awareness of finitude. And, says Tillich, "like finitude, anxiety is an ontological quality. It cannot be derived; it can only be seen and described. *Occasions in which anxiety is aroused must be distinguished from anxiety itself*" (*ST*, I:191). (Italics added.) In *The Courage to Be*, Tillich's definition of Angst has shifted from man's awareness of finitude to his awareness of the *possibility of non-being experienced* in his finitude. Says Tillich:

> . . . anxiety is the state in which a being is aware of its possible nonbeing. The same statement, in a shorter form, would read: anxiety is the existential awareness of nonbeing. "Existential" in this sentence means that it is not the abstract knowledge of nonbeing that produces anxiety but the awareness that nonbeing is part of one's own being. It is not the realization of universal transitoriness, not even the experience of the death of others, but the impression of these events on the always latent awareness of our own having to die that produces anxiety. Anxiety is finitude, experienced as finitude. . . . It is the anxiety of nonbeing, the awareness of one's finitude as finitude. (*CB*, pp. 35–36)

As can be seen from the last sentence in this passage, Tillich does not change his definition of Angst from one work to the other; rather he shifts his emphasis from Angst and the "ontological" analysis of *finitude* in *Systematic Theology* to the "existentialist" analysis of *courage* as an ethical response the threat of non-being.

One other important preliminary observation may be helpful here to distinguish the meanings of Angst in Tillich's two works. In *Systematic Theology*, Tillich says that the "occasions in which anxiety is aroused must be distinguished from anxiety itself" (*ST*, I:191.) The purpose of *The Courage to Be* is to discuss the ontological underpinnings of such occasions, so as to explicitly show how the faith of courage allows man to live life through "in-spite-of" Angst.

Essentially Tillich presents three new facets of his understanding and interpretation of Angst in *The Courage to Be*. All three are expressions of the distinct conditions from which non-being threatens man. As such they are ontological underpinnings to the specific occasions of Angst discussed above. Tillich discusses Angst in the face of: (1) fate and death, (2) emptiness and meaninglessness, and (3) guilt and condemnation. Tillich tells us that the first element of each pair refers to the *relative* or ontic anxiety we experience in life, while the second of each pair is the underlying ontological dimension being threatened in an *absolute* or ontological sense. Thus, the second element more closely approaches the ultimate Angst known, according to Tillich, as despair (*CB*, p. 41).

The first way that nonbeing threatens man, and by far the most devastating in Tillich's own view, is a threatened loss of our "*ontic* self-affirmation" (*CB*, p. 42). This means, according to Tillich ". . . the basic self-affirmation of a being in its simple existence." The first and most universal facet of Angst is revealed in the face of the first "ontic" threat—"the anxiety of fate and death" (*CB*, p. 42). This facet is, of course, a dual one. The relative anxiety in the face of the fate may be distinguished from the absolute Angst apprehended in the face of death. Specifically, Tillich says regarding this first complex facet, "the anxiety of death is the permanent horizon within which the anxiety of fate is at work" (*CB*, p. 43). The term "fate," then, represents an entire group of specific "ontic" anxieties that share one common property: "their contingent character, their unpredictability, the impossibility of showing their meaning and purpose" (*CB*, p. 43). The anxiety of fate is therefore grounded in a full apprehension of my ultimate existential contingency: my utter unnecessariness, and my completely accidental occurrence in being.

On the other hand, ontological Angst, revealed in the face of death, is an absolute threat to man's self-affirmation; death stands behind my ontic contingent being as its end-point and therefore the proof of contingency itself. Death is the ultimate claim against me; it robs me of all final security, sense of belonging to home, of ultimate social and individual existence. Finally, death, says Tillich, ". . . stands behind the attacks on our power of being in body and soul by weakness, disease, and accidents. In all these forms fate actualizes itself, and through them the anxiety of nonbeing takes hold of us" (*CB*, p. 45).

The second facet of ontological Angst is revealed by the threat of nonbeing to man's *spiritual* self-affirmation: relatively, as the "anxiety of emptiness," and absolutely, as the "anxiety of meaninglessness." Specifically, spiritual self-affirmation occurs, Tillich tells us, at every moment of my creative life. I need not be a creative artist to be spiritually creative; rather I must be able to *participate* in culture, participate *actively* in the meanings given to Being by original creations. This is what Tillich means by the spiritual or cultural life in the present context. Hence the ontic form of anxiety as emptiness is revealed when I apprehend a threat to the special contents of my spiritual life. Tillich describes this ontic anxiety as follows:

> A belief breaks down through external events or inner processes: one is cut off from creative participation in a sphere of culture, one feels frustrated about something which one had passionately affirmed, one is driven from devotion to one object to devotion of another and again on to another, because the meaning of each of them vanishes and the creative eros is transformed into indifference or aversion. Everything is tried and nothing satisfies. (*CB*, pp. 47–48)

The absolute Angst associated with the threatened loss of spiritual self-affirmation is Angst in the face of meaninglessness, or in Tillich's words, "the anxiety of meaninglessness is anxiety about the loss of an ultimate concern, of a meaning which gives meaning to all meanings" (*CB*, p. 47). When the anxiety of emptiness becomes more pronounced, it ". . . drives us to the abyss of meaninglessness" (*CB*, p. 48). At the outset of this absolute

Angst, I feel at first a grave doubt about my spiritual center, becoming cognitively separated from my spiritual values. The threat to this spiritual realm, says Tillich, is not simply some enhanced form of discriminatory doubt, the doubt necessary for making the aesthetic judgments; rather, Tillich means the threat takes the form of *total doubt*—the doubt that leads to despair. To attempt to combat the rapid onrush of such Angst, man may cling to traditions and accept the quiet desperation of resignation. Then, if this does not dispel the despair (as surely it cannot), we next relinquish our spiritual freedom—the freedom to ask questions and doubt answers. Specifically man ". . . surrenders himself in order to save his spiritual life. He escapes from his freedom (Fromm)/100/ in order to escape the anxiety of meaninglessness" (*CB*, p. 49). In this way meaning is saved, says Tillich, but the self is sacrificed.

Finally, the third facet of primordial Angst arises as the threat of nonbeing to man's *moral* self-affirmation. Relatively, this is apprehended as the anxiety of guilt. Absolutely, however, Angst in the face of ultimate condemnation threatens man's moral self-affirmation. Specifically, Tillich states, "man's being, ontic as well as spiritual, is not only given to him but also demanded of him" (*CB*, p. 51). This means I am responsible for my being and must be required to answer for what I have made of my life. Each of us must ask this question, says Tillich; and likewise each of us must serve as prosecutor and judge in the court of our individual moral self-affirmation.

From the ontic or relative standpoint, the defendant-prosecutor-judge function gives rise to the anxiety of guilt. "Man," says Tillich (in direct opposition to Sartre), "is essentially 'finite freedom' . . . free within the contingencies of his finitude. But within these limits he is asked to make of himself what he is supposed to become, to fulfill his destiny" (*CB*, p. 52). Insofar as I *am* finitude, I cannot become perfect. Indeed, even my *best deed* is prevented from being perfect by the radical finitude at my essence. Hence, as judge at the bar of personal responsibility, I must ask myself to account for what I have become: I must experience my own radical inability to become perfect. The apprehension of this ontic fact is, says Tillich, the anxiety of guilt.

Behind this ontic anxiety there stands the Angst of total condemnation: the possibility of being condemned to the despair of having lost my worth as a human being and my destiny as a person. To compensate for such ultimate Angst, I can either defy all negative judgments about my Self, as well as the moral demands upon which they are based (the position of animism), or I can flee into moral rigor with *its* satisfactions (the position of legalism). But in either case, moral Angst abides, waiting to break into the open ". . . producing the extreme situation of moral despair" (*CB*, p. 53).

Tillich next takes up the nature of *pathological* anxiety. We mention his discussion only to distinguish purely psychological or neurotic anxiety from ontological Angst. Tillich begins his discussion of pathological anxiety by observing that "non-existential anxiety, which is the result of contingent occurrences in human life, has been mentioned only in passing" (*CB*, p. 645), but now the time has come to take up "non-existential anxiety" both explicitly and systematically. Tillich observes that the several psychotherapeutic theories of neurotic anxiety have one common denominator: they see anxiety as ". . . the awareness of unsolved conflicts between structural elements of the personality . . ." (*CB*, p. 64). These conflicts are between unconscious drives and repressive norms, for example, or between imaginary versus real worlds. Yet the basic problem systemic to psychotherapeutic theory *in general* is that its theoreticians as well as its practitioners *cannot agree upon what is basic to the nature of neurotic anxiety as opposed to what is derived in the form of symptoms.* The result is utter confusion. Says Tillich:

> It is the lack of a clear distinction between existential and pathological anxiety, and between the main forms of existential anxiety. *This cannot be made by depth-psychological analysis alone*; it is a matter of ontology. Only in the light of an ontological understanding of human nature can the body of material provided by psychology and sociology be organized into a consistent and comprehensive theory of anxiety. (*CB*, p. 65) (Italics added.)

Thus, pathological anxiety is Angst under special *psychological conditions.* What these conditions are, are dependent upon the various relationships between self-affirmation, courage, and Angst. Tillich outlines the dynamics of these relationships on the ontological and moral level, but ultimately the moral level triumphs: "Courage resists despair by taking anxiety into itself"

(*CB*, p. 66). This clue concerning the nature of courage makes possible a deeper understanding of pathological anxiety. When man becomes incapable of self-affirmation "in-spite-of" despair, he succeeds ". . . in avoiding the extreme situation of despair by escaping into neurosis" (*CB*, p. 66). We see, then, that neurosis, ". . . *the way of avoiding nonbeing by avoiding being*" (*CB*, p. 66), is the very epicenter of the ontic phenomenon of pathological or non-existential anxiety. Primordial Angst, as the rock bottom ontological condition of human finitude, is never brought to the surface. It is shortcircuited by neurosis.

Preliminary Conclusions Regarding Tillich and Angst

We may now draw to a close our brief outline of how primordial Angst is revealed in Tillich's thought by raising this question: What have we learned here that furthers Angst's revelation in philosophical thought? Keeping in mind that contrasts and comparisons are very close at hand in the following chapter, we shall restrict our comments at this point to little more than a skeletal sketch of the promises and problems engendered from Tillich's contribution.

First, regarding Tillich's promises, we observe the following six points: (1) Tillich's analysis of Angst is the radical synthesis of both Heidegger's and Kierkegaard's analyses. This synthesis makes possible an active solution to "the problem" of Angst—namely, courage. (2) Angst in Tillich's view is both an ontological phenomenon and an ethical one. So Angst provides a connection point between phenomenological ontology and existential philosophy. (3) Tillich's conception of Angst as a consequence of man's finite freedom displays a positive role for Angst, namely, as what makes courage and the joy of self-affirmation possible in the face of the abysses of death, meaninglessness, and condemnation. (4) Tillich's analysis of Angst as the "ontological shock" in the face of the abyss is the phenomenological linkage between Heidegger's atheism and Otto's theism. This makes the *mysterium tremendum et fascinosum* an integral part of his Angst analysis. (5) Tillich's analyses of the many facets of primordial Angst makes possible a synthetic apperception of what he calls "ontological anxiety" as the *ultimate ground* for Angst in the face of guilt and condemnation, and Angst in the face of emptiness and meaninglessness.

But over against this important series of contributions to

revealing Angst's meaning, there stands a set of provisional problems that will demand attention in the following chapter. We may roughly sketch these problems as follows.

1. Tillich holds that man is *more* than Dasein. Man has a Self that possesses a conscious, a subconscious, and an unconscious. This assumption may in the end provide the downfall of Tillich's ontological analysis of Angst, for it opens his analysis to the charge that after all Angst is not an ontological phenomenon, but rather one springing from the wholly immanent abyss of the human unconsciousness, a view generally argued by the depth-psychologists. For reasons we shall discuss later, we cannot accept this view. The admission of man's ego as "possessing" these psychic attributes clearly opens his analysis up to an immanentist attack.

2. Tillich tells us that Angst is at once the apprehension of our finitude as a matter "internal" to man, and the apprehension of the threat of non-being as an "external" matter. While his analysis of finitude is made clear, the notion of "non-being" is never clarified except as that which conditions the threat to man's self-affirmation. Ultimately "non-being" remains a mystery, a meontic nothing that stands in dialectical tension over against Being. So if Sartre errs on the side of *too complete* a description of Nothingness, Tillich may well err on the side of *incompleteness*—a charge that makes his position vulnerable to the more drastic accusations of obscurantism regarding the ontological character of Angst as the threat of non-being.

3. Finally, there is the problem of the so-called *answer* to Angst, the phenomenon of courage. If courage is a phenomenological phenomenon, then we must be able to describe precisely its elements both ontically and ontologically. Tillich does this ontically throughout both his *Systematic Theology* and *The Courage to Be*. But the ontological condition of the possibility of courage is primordial Angst in its various facets of death, meaninglessness, and condemnation. So if courage is to be a genuine answer to the challenge of Angst it must have its ontological ground in something other than Angst. This is, of course, the absolute faith in man's ultimate concern which Tillich calls the God beyond God; and the justification for the term "meta-theology" for his theonomous philosophy. This "answer," even though Tillich tells us that it is a *necessary* one, was clearly not necessary to Sartre, for example, who would argue that Tillich's

answer is no answer at all. In fact it is little more than an elaborate cover-up of man's *de trop* being in a universe without God or the God above God. In other words, Tillich is here open to the charge of confusing his *experience* of faith with the ultimate *justification* for faith, a charge against which there can be no sound phenomenological or logical defense.

It is not necessary to belabor our point further. There are obvious strengths and weaknesses on both sides of Tillich's conception of Angst. While we would never presume to determine the final adequacy of Tillich's portrait of Angst, we would suggest that such a judgment may turn upon Tillich's "beyond theism" or "meta-theism" that seeks to answer the antimonies raised by existential analysis of the human condition. Such is the whole point, after all, of Tillich's correlative approach to theonomous philosophy. The adequacy of both Tillich's analysis of ontological Angst and ethical courage may be confirmed only if the statement representing the culmination of his analysis is existentially true, even if it is not phenomenologically adequate: namely, "the courage to be is rooted in the God who appears when God has disappeared in the anxiety of doubt" (*CB*, p. 190). If such a statement is false regarding either existential truth or phenomenological adequacy, then the argument regarding the dialectial structures of Angst and courage becomes a purely theological one—not one which carries the force of phenomenological philosophy, a force for which Tillich clearly strives.

NOTES

/1/ Alexander Astruc and Michel Contact, *Sartre by Himself*, translated by Richard Seaver (New York: Urizon Books, Inc., 1978), p. 50. Note: Sartre tells us that during his captivity he had read Heidegger a second time, and ". . . three times a week I used to explain to my priest friends Heidegger's philosophy." On this same page it is pointed out that according to Simone de Beauvior, Sartre returned from captivity as a fundamentally changed person—one who had become rigid with moral righteousness.

/2/ Rollo May, *Paulus: Reminiscences of a Friendship* (New York: Harper & Row, 1973), p. 18. Note: "Paulus" was the name Paul Tillich was known by in his native German, according to May. Cf. p. 1 of the above cited work.

/3/ J. Glenn Gray, *The Warriors: Reflections of Men in Battle* (New York: Harcourt Brace, 1959).

/4/ Which means, of course, that the present chapter might be somewhat lengthy vis-à-vis the three which precede it. But since this work is concerned with twentieth-century philosophical thought and the disclosure of Angst within it, one long chapter that makes contrasts and comparisons possible between the theistic and atheistic modes of Angst's disclosure is in our view superior to two chapters in which this flow of contrasts and comparisons is broken up.

/5/ Jean Paul Sartre, "La Transcendence de L'Ego: Esquisse d'une description phenomenologique." *Recherches Philosophiques*, VI (1936–37), trans. and annotated by Forest Williams and Robert Kirkpatrick as *The Transcendence of the Ego: An Existentialist Theory of Consciousness* (New York: The Noonday Press, 1957).

/6/ This is, of course, the proper translation into English of the subtitle of Sartre's work. At the time of this work Sartre could not possibly have agreed to the term "Existentialist" in this title. Besides this, the French clearly states that it is an "outline" (*Esquisse*) of a *phenomenological* description.

/7/ Sartre, "Existentialism," p. 36. Cf. Kaufmann, *Existentialism from Dostoevski to Sartre*, p. 302.

/8/ Jean Paul Sartre, *Les Mots* (Paris: Librairie Gallimand, 1964), trans. by Bernard Frechtman as *The Words* (New York: Vintage Books, 1981), p. 251.

/9/ Marjorie Grene, *Sartre* (New York: New Viewpoints, 1973), p. 39.

/10/ *Sartre by Himself* was originally a film-length interview in which many persons participated, posing questions to Sartre. The film was begun in 1972 and, after a three year hiatus, was first shown at the Cannes Festival on May 27, 1976.

/11/ Ibid., p. 25.

/12/ Ibid., p. 29. Note: the text points out that Sartre meant to say "Levinas" rather than "Gurvitch."

/13/ Ibid., pp. 29–30.

/14/ *BN*, "translator's introduction," p. xiii.

/15/ Sartre, *The Transcendence of the Ego*, p. 21.

/16/ Ibid., p. 22.

/17/ Speigelberg, *The Phenomenological Movement*, II:451–52.

/18/ Ibid., p. 463. Note: Speigelberg relates that Heidegger did not remember meeting Sartre prior to Sartre's Freiburg lecture of 1953. Yet the following account of Sartre's early meetings with Heidegger is perhaps worth repeating. Speigelberg says: "When asked soon after the war about his early acquaintance with Sartre, Heidegger did not remember him by name; then he identified him as 'the Frenchman who had always confused him with Husserl.' Sartre's primary interest at that time was clearly in Husserl. It was not until the period of *L'Être et le néant* that he became more keenly interested in Heidegger's own philosophy. His reaction to Heidegger personally was apparently negative. Thus in commenting on Heidegger's political role, he stated publicly: 'Heidegger n'a pas de caractère. Voilá la vérité' (*Action*, December 27, 1944; *Lettres*, Geneve, I [1945], p. 83). Nevertheless, Sartre was one of the first to intercede for Heidegger after the French occupation of Freiburg to the extent of wanting him to be invited to Paris." For those who are interested, the translation of Sartre's comment is something like, "Heidegger does not have any character. That is the truth." While this is not the place to discuss Heidegger's Nazi connections, we feel compelled at least to state that in our view there can be no excuse for this event in his life as a philosopher. Some of the documents that have recently come to light have clearly shown that his involvement was more than marginal. Cf. Walter Kaufmann, *The Discovery of Mind*, II:219–24, for some of the more glaring examples of this involvement.

/19/ Jean Paul Sartre, *Esquisse d'une theorie des emotions*, trans. by Bernard Frechtman, *Outline of a Theory of the Emotions* (New York: Philosophical Library, 1948) and by Philip Mairet, *Sketch for a Theory of the Emotions* (London: Methuen, 1962).

/20/ Speigelberg, *The Phenomenological Movement*, II:453.

/21/ Ibid.

/22/ Ibid., p. 454.

/23/ Ibid.

/24/ Ibid.

/25/ *Sartre by Himself*, p. 25.

/26/ Jean Paul Sartre, *Critique de la Raison Dialectique*, translated in part by Hazel Barnes, *Search for a Method* (New York: Knopf, 1963).

/27/ *BN*, pp. 45–47.

/28/ *BN*, "translator's introduction," p. xxii, footnote 9.

/29/ William Leon McBride, "Man, Freedom and *Praxis*," *Existential Philosophers: Kierkegaard to Merleau-Ponty*, edited by George A. Schrader, Jr. (New York: McGraw-Hill, 1967), p. 271, footnote 13.

/30/ Jean Wahl, *Jules Lequier* (Paris: Editions des Trois Collins, 1948), as cited in McBride, "Man, Freedom and *Praxis*," p. 271, footnote 13.

/31/ There is a possibility, of course, that Lequier could have been cited in one of any of Sartre's unpublished works, such as, for example, the one mentioned in *Sartre by Himself* entitled "Psyche," which was not completed. Cf. *Sartre by Himself*, p. 50.

/32/ Speigelberg, *The Phenomenological Movement*, II:469.

/33/ *Sartre by Himself*, pp. 50–51.

/34/ *BN*, "translator's introduction," p. xxiv.

/35/ Spiegelberg, *The Phenomenological Movement*, II:453.

/36/ Space considerations prevent us from any treatment of "bad faith" beyond its mention in this chapter. We recognize, however, that any full-dress treatment of the phenomenon of Angst in Sartre's thought would require an analysis of "bad faith" to display the means whereby we seek to overcome essential Angst interpreted of course as *angoisse*.

/37/ *BN*, "translator's introduction," p. xxii. Cf. *BN*, p. 4.

/38/ Heidegger, *WM*, p. 11/*G*, p. 114.

/39/ The use of the first person is, of course, the phenomenological "I" which examines what is given to it in consciousness. This is a precedent, established by Husserl at the beginning of his own phenomenological investigations. We shall use it throughout out the remainder of this essay.

/40/ Kaufmann, *Existentialism from Dostoevski to Sartre*, p. 45.

/41/ Speigelberg, *The Phenomenological Movement*, II:473.

/42/ *Sartre by Himself*, pp. 74–75.

/43/ Sartre, "Existentialism is a Humanism," in Kaufmann's translation, pp. 290–91.

/44/ Ibid., p. 291.

/45/ Ibid.

/46/ Ibid., p. 293.

/47/ Ibid.

/48/ Ibid., p. 294.

/49/ Ibid.

/50/ *BN*, "translator's introduction," p. xxxiv.

/51/ Ibid.

/52/ Rollo May, *Paulus* (New York: Harper & Row, 1973), p. 71.

/53/ Ibid., p. 29.

/54/ The reader will recall that in discussing how May uncritically accepted Tillich's insistence that Angst should be translated as "anxiety" rather than as "dread" in Lowrie's translation of *Begrebnet Angst*, we argued that May had violated a principal canon of hermeneutic phenomenology: namely, the appeal to Tillich as an authoritative source, a circumstance which Heidegger saw as an external signification placed over the things themselves which is grounded upon "fancies or popular conceptions." Cf. *BT*, p. 195.

/55/ David Hopper, *Tillich: A Theological Portrait* (Philadelphia and New York: J. P. Lippincott Company, 1968). (Hereafter, *Tillich*.)

/56/ Charles W. Kegley and Robert W. Bretall, *The Theology of Paul Tillich* (New York: The Macmillan Company, 1952), p. 9.

/57/ Ibid., p. 8.

/58/ Paul Tillich, *On the Boundary: An Autobiographical Sketch* (New York: Charles Scribner's Sons, 1966), pp. 46–47.

/59/ Tillich read the complete works of *Schelling* twice before presuming to write his dissertation(s) on Schelling. Cf. Tillich, *On the Boundary*, p. 47.

/60/ This work has been translated into English by Victor Nuovo as *The Construction of the History of Religion in Schelling's Postive Philosophy* (Lewisburg, PA: Bucknell University Press, 1974).

/61/ This work has likewise been translated by Victor Nuovo as *Mysticism and Guilt Consciousness in Schelling's Philosophic* Development (Lewisburg, PA: Bucknell University Press, 1974).

/62/ Ibid., "translator's introduction," pp. 9–10.

/63/ Ibid.

/64/ Other antimonies discussed by Tillich in this connection are those between identity and difference, unity and the manifold, and theoretical and practical reason. Cf. Tillich, *Mysticism and Guilt Consciousness*, "translator's introduction," pp. 11–12.

/65/ Ibid.

/66/ Bernard Martin, *The Existentialist Theology of Paul Tillich* (New York: Bookman Associates, 1963), p. 18.

/67/ Ibid.

/68/ Ibid.

/69/ Tillich, *On the Boundary*, p. 53.

/70/ Ibid., p. 48.

/71/ Ibid., pp. 48–49.

/72/ Ibid.

/73/ When he met Heidegger, then a young associate Professor at Marburg, Tillich was himself a visiting associate professor of Theology.

/74/ Kegley and Bretall, *The Theology of Paul Tillich*, p. 11.

/75/ Paul Tillich, *My Search for Absolutes* (New York: Simon and Schuster, 1967), p. 42.

/76/ Hannah Tillich, *From Time to Time* (New York: Stein and Day, 1973), p. 117.

/77/ "Paulus," as May's book on Tillich tells us, was Tillich's "familiar" name. Cf. May, *Paulus*, p. 1.

/78/ Hannah Tillich, *From Time to Time*, p. 117.

/79/ Kegley and Bretall, *The Theology of Paul Tillich*, p. 14.

/80/ Tillich, *On the Boundary*, p. 57.

/81/ Ibid., p. 56.

/82/ Ibid., pp. 56–57.

/83/ Ibid., p. 54.

/84/ Hannah Tillich, *From Time to Time*, p. 117.

/85/ Kegley and Bretall, *The Theology of Paul Tillich*, p. 6. Note: for Otto's work cf. Rudolf Otto, *Das Heilige*, trans. by John W. Harvey as *The Idea of the Holy: An Inquiry into the Non-rational Factor in the Idea of the Divine and its Relation to the Rational* (London: Oxford University Press, 1958).

/86/ Kegley and Bretall, *The Theology of Paul Tillich*, p. 6.

/87/ Tillich, *My Search for Absolutes*, p. 129.

/88/ Ibid., p. 130.

/89/ *ST*, I:216.

/90/ Otto, *The Idea of the Holy*, p. 15. Note: it is interesting that Otto employs a full range of possible terms to describe the subtle nuances of the *tremendum* experience. He mentions the Hebrew *higdish* (hallow), meaning "keep a thing holy in the heart," for example, to mark off a feeling of dread "not to be mistaken for any ordinary dread," but rather one that is exclusively related to the experience of the numinous (p. 15). But moreover Otto likens the *mysterium* feeling of *tremendum* to *te'emah* or fear of Yahweh; to the *augustus* of the Greeks, a name given only properly to the numen despite the Roman corruption of it; to several German words such as *erschauern* (to quiver with emotion), and the baser German concepts of *grausen* (to have a horror of), *grasslich* (grisly), and specifically *Scheu* (dread). Finally, in English Otto suggests the words "awe" and "aweful" ". . . in their deeper and most special sense approximate closely our meaning" (p. 14). Needless to say, this etymological excursion of the religious phenomenon of *mysterium tremendum* is likewise close to our own analysis of the self-revelation of primordial Angst as an ontological phenomenon discussed by us in Chapter I of the present work.

/91/ This capsule summary excludes the important influences of Hegel, Marx, Hursch, Barth, and a host of others that we cannot consider here.

/92/ *ST*, I:191.

/93/ *CB*, p. 35.

/94/ The vision of the unification and bifurcation of philosophy and theology was an ever present problem for Tillich. As early as 1923, in his work *Das System der Wissenschaften nach Gegenstanden und*

Methoden [*The System of Science According to its Objects and Methods*] (Goettingen: Vandenhoeck & Ruprecht, 1923), Tillich described and explained his concept of theology as "theonomous philosophy" or "theonomous metaphysics." Philosophy per se, says Tillich, is concerned with the finite and the conditioned, regarding the unconditioned as ground. Theonomous philosophy, on the other hand, "turns toward the unconditional for its own sake, using the conditioned forms to grasp the unconditional through them" (Martin, *The Existentialist Philosophy of Paul Tillich*, p. 30). Hence although philosophy and theology as theonomous philosophy differ, the highest of the latter is "to pass beyond its own independence, bringing to expression its unity with autonomous philosophy" (ibid.).

/95/ Martin, *The Existentialist Theology of Paul Tillich*, p. 218.

/96/ The concept of "ultimate concern" is, of course, the phenomenological description associated with God. Tillich says, "'God,' is the answer to the question implied in man's finitude: he is the name for that which concerns man ultimately" (*ST*, I:211). It is very important to realize that this phenomenological description takes its point of departure from Otto's experience of the Holy. Tillich therefore cements the phenomenological nexus between himself and Otto by asserting without equivocation: "Only that which is holy can give man ultimate concern, and only that which gives man ultimate concern has the quality of holiness" (*ST*, I:215).

/97/ The "New Being" refers to "the quest for a new and saving power that will heal the disruptions and estrangements between an old religion and culture and a new one." Christianity's assertion is that the New Being "manifests itself in a unique and dynamic way" in the person of Jesus of Nazareth. Cf. Wayne W. Mahjan, *Tillich's System* (San Antonio, TX: Trinity University Press, 1974), pp. 32–33, for an excellent account of this concept in Tillich's thought beyond that given in *ST*, II:78–96.

/98/ Unlike Heidegger and Sartre, Tillich accepted as relevant to *Systematic Theology*, at least, the insights of depth-psychology. Thus, the unconscious is related to the "dynamics" element of the "dynamics-form" polarity as but one manifestation of this ontological pole to which correspond the "*Urgrund*" of Boehme, the "will" of Schopenhauer, the "will-to-power" of Nietzsche, the "*elan vital*" of Bergson, and the "strife" of Jung. But Tillich extends his unconsciousness notion not only to Freud but to Hartmann as well, seeing it as ". . . mere potentiality" (*ST*, I:179).

/99/ It is very likely the contents of this footnote that provided the support for Rollo May's statement in *The Meaning of Anxiety* where he cites the authority of Tillich in asserting that what Kierkegaard meant by Angst was "anxiety" rather than Walter Lowrie's translation as "dread." Cf. May, *The Meaning of Anxiety*, pp. 36–37, note 39.

/100/ Erich Fromm, *Escape from Freedom* (New York: Avon Books, 1965).

CHAPTER V

THE TOPOS OF ANGST IN THE WESTERN TRADITION

The Question at Hand

We must, however, now ask: is there a broader hermeneutical horizon wherein the essence of Angst can be grasped? More specifically, is there a hermeneutical synthesis in which the multiplicity of these many portraits of Angst can be brought together? Is there, in other words, a broader vista of Angst which has escaped our attention thus far due to the detailed micro-hermeneutical approach of the preceding four chapters?

To these questions we unequivocally respond positively: there *is* a broader horizon for grasping the essence of Angst. The task of the present section is (1) to reveal the phenomenological parameters of this horizon and (2) to sketch out Angst's topography in hermeneutical terms. In this manner we will approach our final interpretation of Angst's role in Western thought with a clear dose of measured caution. For to announce boldly such an interpretation without preparing the ground will defeat what has heretofore been a careful and systematic attempt to prevent "fancies and popular conceptions" from obscuring a clear and final vision of primoridal Angst.

Thus our first task must be to evaluate critically the adequacy of the descriptions of primordial Angst provided in the chapters above. Only afterwards can we address contrasts and comparisons of Angst at the macro-hermeneutic level.

A Point of Departure—the Phenomenon of Apprehension

Our basic clue to understanding primordial Angst's broader horizon lies in the linguistic expression we have given in describing "primordial Angst": namely, "the pre-reflective apprehension" of the abyss separating finite being from infinite Being, the temporal from the eternal, the contingent from the necessary, or the created from the creator. Two key terms now demand closer examination, namely: "apprehension" and "pre-reflective."

We have used the term "apprehension" throughout this essay to pin-point, so to speak, a general pre-sentiment in Dasein associated with the Angst phenomenon. Let us now try to make this pre-sentiment more specific. "Apprehension" has, of course, several distinct but related meanings. Among those offered in the *Oxford English Dictionary*: "fear as to what might happen; dread."/1/ We have attempted in the present essay to refine this definition by pointing out, for example, the strengths and weaknesses of the terms "fear" and "dread" in relation to primordial Angst. Each of the thinkers we discussed here, Kierkegaard, Heidegger, Sartre, and Tillich, agreed: fear is reserved for direct threats to man's being. But "apprehension" as defined above by the *Oxford English Dictionary* specifies the "fear" of "what might happen" in the future. Thus the futural direction of apprehension is explicitly recognized. Such a characterization of apprehension's futural dimension would meet the descriptions of Angst provided by Kierkegaard and Heidegger, but it falls short of those provided by Sartre and Tillich. Specifically, for both Kierkegaard and Heidegger Angst's primary temporal direction appears to be the future as we saw in Chapters II and III. But for Sartre and Tillich Angst spans the temporal directions of past, present, and future. Angst of the future is related to Angst of the past and both come together in the existential "now." Moreover, for the "existentialist" writers Angst is clearly related to a *Self*. Thus its primary temporal significance is likewise couched in the "now" of a Self making choices (for Sartre) and the "now" of self-affirmation in spite of threats to my non-being (Tillich). This is not to say that Sartre and Tillich deny the futural dimension of Angst. They simply do not emphasize it to the same degree as did Kierkegaard and Heidegger, who both look to man's *telos* as being futural.

A second definition of "apprehension" has enormous ontological possibilities; for this definition is precisely in line with Heidegger's insistence that Angst, as Dasein's basic pre-disposition, cannot be separated from understanding and speech as equiprimordial modes of Dasein's being-in-the-world. Specifically, "apprehension" also means: "the apprehensive facility; the ability to understand; understanding."/2/

As we have seen, each thinker we have considered at the

micro hermeneutical level either explicitly or implicitly accepts the relationship between Angst and understanding. In Kierkegaard, for example, Angst as "sympathetic antipathy and antipathetic sympathy" is what makes understanding possible for the subjective individual. Angst conditions the subjective individual's transcendence from one mode of existence to another; that is, he cannot grasp a new existence medium until he takes a leap originating in the Angst of the "dreaming spirit." Heidegger explicitly recognized the necessary relationship between Angst and understanding as two dimensions of the complex Being-in-the-world phenomenon. Sartre sees the *angoisse* of freedom as the necessary condition for understanding; *réalité humaine*'s mode of transcending the *en-soi* via the Nothingness that defines consciousness *is* understanding for Sartre. Tillich openly recognizes Angst as the explicit understanding we each have of our finitude. These facts should provide yet another major insight into Angst. Angst is a pre-reflective, non-discursive, pre-understanding, or pre-dispositional attunement that makes possible discursive understanding in the sense that the term "understanding" is ordinarily used.

If our understanding and interpretation of "apprehension" is correct, and if, moreover, apprehension corresponds closely to what Heidegger meant by Angst, then have we not simply reduced primordial Angst to "apprehension"? Furthermore, have we not done precisely what we initially said could not be done, namely, to provide an adequate English translation for Angst?

Apprehension: A Closer Look

While these questions appear to undermine our efforts here, this need not be so provided we move to a deeper phenomenological analysis regarding "apprehension." The term "apprehension" is wholly inadequate phenomenologically, we submit, to display the plethora of meaning inherent in primordial Angst. For according to our description of primordial Angst, along with "apprehension" goes the decisive modifier "pre-reflective"—a term whose full meaning is revealed only through a phenomenological analysis and description. And, as we shall see, without this modifier, primordial Angst cannot be distinguished from the ontic terms "anxiety," pathological "anxiety," "fear," "worry," etc.

But to assert glibly or naively that primordial Angst is a pre-reflective mode of apprehension is likewise wholly inadequate without further elaboration of the *sense* in which "pre-reflective" is used. Clearly, in one sense "pre-reflective" must mean something like "prior to reflection." Hence, we may find it useful to grasp the meaning of "reflection" in connection with Angst before attempting to get beneath the reflective level to a more primordial level called "pre-reflective."

That to which both "reflective" and "pre-reflective" point must be "consciousness" or my opening to Being (as Dasein): the very essence of both Husserl's transcendental phenomenology and Sartre's phenomenological ontology. While there is neither space nor point to elaborate the history of the consciousness concept here, we feel compelled to observe at least the following: the apodicticity of the Cartesian *cogito*, when coupled with the notion of a *res cogitans* or "a subject which thinks," contains the full-dress potentiality for Husserl's transcendental phenomenology.

Kant, in the *Critique of Pure Reason*, was very much concerned with the problem of the limits of consciousness, or what we can truly "know" on the basis of our finite understanding. Husserl, of course, saw all consciousness as intentional; that is, consciousness is an activity distinct from the so-called "mental objects" *of* consciousness. For phenomenology, as Husserl interprets that discipline, the apodictic *cogito* must never be confused with the *cognitationes*. Accordingly, the objects of consciousness are always intended *by* consciousness itself. An implied corollary to this intentionalist view is this: even when consciousness becomes reflexive, that is, when it refers back to itself as the source of consciousness, self-consciousness can always become a theme for phenomenological investigation. Why? Precisely because self-consciousness is constituted as an intentional act. But in examining self-consciousness it is necessary to bracket out all existential considerations of value, of actual existence, of history, and of circumstance to get to the thing-itself, which *in* itself must be studied *as* an intended act of consciousness. This bracketing out, the formal *epoché* of Husserl, is the necessary standpoint to successfully carry out any "transcendental" or "pure" phenomenological investigations.

In phenomenologically investigating consciousness as self-consciousness (a kind of meta-phenomenological inquiry), the question naturally arises as to whether or not an "ego" exists *behind* and under the activity of consciousness. Clearly, Husserl affirms this ego as a transcendental ego, an "I" carrying out the activity of consciousness. But this assumption, at least in the view of both Heidegger and Sartre, must result in the very subject-object dichotomy Husserl's early phenomenology sought to undercut: it re-introduces the ego as the medium through which the things themselves must be studied. Thus the "ego" becomes a subject standing over against the objects it studies.

We can overcome this problem, as well as many others raised by the notion of a transcendental ego, if we employ Ockham's razor with a new phenomenological edge to cut out the notion of an ego abiding behind the intentionality of consciousness. So in this new view the notion of "ego" or "Self" becomes simply another object of consciousness, intended by consciousness as among the things of the world: a kind of being amidst other beings. For Sartre the consequence of this view was that no "thing" or "ego" constitutes the objects of consciousness into intended objects: as we saw in Chapter IV, "consciousness is a great emptiness, a wind blowing toward objects."/3/ Consciousness *is* Nothingness, but a Nothingness never isolated from *cognitationes* of pure consciousness. Heidegger completely rejects the notion of consciousness as standing over against the world of intentional objects, and in particular he discards any notion of a transcendental ego. Rather, Dasein is a being already there in a world of circumspect concern, a world where meanings are referential in the *context* of world. The notion of transcendental consciousness is likewise rejected by Tillich/4/ on the grounds that while Husserl's phenomenology is "competent in the realm of logical meanings . . . it is only partially competent in the realm of spiritual realities like religion."/5/

All of this is, of course, post-Husserlian. Thus the problems of consciousness, transcendental ego, and intended objects could never have occurred to Kierkegaard, who predated Husserl and who was in the grip of Hegelian metaphysics. Thus, Kierkegaard's cognitive approach to method was not phenomenological per se, but rather one couched in his "know thyself!" principle, along with the *unum noris omnes* method of experimental and psychological verification—all in the name of man's absolute *telos*, the eternity of Christian existence. Yet prior to the Self's emergence in despair, prior to the leap into the abyss that takes its launching point from Angst, the spirit is "dreaming"; that is, the spirit is vividly conscious but not as yet *self-conscious* of the Nothingness surrounding it as it approaches the abyss. Thus in this limited sense, consciousness is intentional for Kierkegaard as well.

The Two Approaches to "Apprehension"

This brief description is meant to serve as background for our main discussion of "reflective" versus "pre-reflective" Angst. In "reflective Angst" the *angoisse* of Sartre and the "ontological anxiety" of Tillich, a Self is interlarded between spontaneous awareness of the abyss and the daunting and alluring threat of the abyss itself. Thus the basic distinction between pre-reflective Angst and reflective Angst is revealed in the spontaneity of pre-reflective Angst as opposed to the Self-reflexivity of reflective Angst. In pre-reflective Angst there is no Self. In reflective Angst there is. Pre-reflective Angst is bound by nothing; reflective is bound by Self. Thus in Sartre and Tillich's "existential" writings, what is threatened in Angst is the "Self." But in Kierkegaard, there is no self-conscious Self to be threatened at the face of the abyss. Rather the Self emerges only *after* the "dreaming spirit" takes the leap into the abyss. Kierkegaard's notion of Angst, then, is originally *pre-reflective*. Angst is spontaneous, that is, bound by Nothingness. Only after the Self emerges from the abyss as a being conscious of its own guilt can the Self become infested with despair and melancholy. Such a condition leads the despair-filled Self to the ultimate leap—the leap of faith. In Heidegger Angst is first and foremost a pre-reflective phenomenon: a pre-reflective or non-self-conscious *fleeing* in the face of my own being-in-the-world. Only within this spontaneous fleeing can the Nothingness of the world's "totality of all possible purposes" [*Bewandtnisganzheit*] be revealed. No Self, then, stands between Angst and the Nothingness of world, although a Self is precisely what is chosen from within the existential *epoché* of Angst. Once we become suspended in Angst's power to make us *solus ipse*, the "being free for (*Freisein für* . . .). Dasein chooses its authentic Self as a mode of being" (*SZ*, p. 188). So Heidegger's notion of Self as authentic being-in-the-world emerges as a result of Dasein's pre-reflective apprehension of Angst. Thus, my chosen Self is authentic and free to become my potentialities.

Likewise in Tillich's thought, Angst is at least initially a pre-reflective apprehension of man's ontological finitude. Only on the basis of this pre-reflective apprehension is anything like reflective Angst possible. But like Sartre, Tillich sees Angst as the reflective awareness of the threat of non-being to human being. Angst is for Tillich ". . . the self-awareness of the finite self as finite."

We find, therefore, two modes of Angst's revelation in the literature we have investigated at the micro-hermeneutical level. On the one hand, primordial Angst is the pre-reflective apprehension of the abyss between the finite and the infinite in Kierkegaard and Heidegger. On the other hand, in Sartre's and Tillich's view, primordial Angst is a reflective apprehension, one related to a Self standing at the abyss sensing a danger to its selfhood.

If our interpretation so far is correct, we must distinguish between these two approaches to Angst in the Western tradition. Reflective Angst, or the Angst of an interlarded Self that *sees* itself at the abyss, must be distinguished from pre-reflective Angst, or the spontaneous Angst in the face of the abyss itself. The schematization for the former appears to be Angst in the face of human freedom. So then let us further deepen our analyses by describing the reflective Angst of Sartre and Tillich.

Reflective Angst and the Self

Angst (in the face of the Self's absolute freedom) may be characterized as a reflective apprehension of my "Self" as a being who must freely choose to actualize my own possibilities in my lifetime.

If such Angst is to be in any way related to primordial Angst, it must possess both the daunting and fascinating descriptive characteristics belonging essentially to primordial Angst. To display such Angst we might find it more illuminating to choose a description not previously provided by our four major philosophers. Rather, let us turn to a representative of the Western tradition who stands historically between Boehme and Kierkegaard, the great enemy of Cartesian rationalism, Blaise Pascal. In his *Pensees*, Pascal observes:

> I know not who put me into the world, nor what the world is, not what I myself am. I am in terrible ignorance of everything. . . . I see those frightful spaces of the universe which surround me, and I find myself tied

> to one corner of this vast expanse, without knowing why I am put in this place rather than another, nor why the short time which is given me to live is assigned to me at this point rather than another of the whole eternity which was before me or which shall come after me. I see nothing but infinity on all sides, which surround me as an atom, and as a shadow which endures only for an instant and returns no more. All I know is that I must soon die, but what I know least is this very death which I cannot escape./6/

The power with which Pascal describes both the daunting and fascinating elements of Angst in this passage by far surpasses many modern descriptions of reflective Angst. And to be sure, parallels exist between Pascal's vision and the "existentialist" interpretations of Sartre and Tillich we have presented at the micro-hermeneutical level. To depict these we suggest the following basic elements of reflective Angst (in the face of the Self's absolute freedom): (1) The insight that freedom of choice *is* what it means to be human shows vividly that human being must be markedly different from any other kind of being. To be human is to choose, and to choose is to transcend the realm of possibilities. Only human being, in other words, *can* opt for one existential possibility over another. Thus my human being is given to me as wholly transcendent. My transcendence means I alone must choose: I am isolated and condemned to my freedom. (2) The full responsibility for my choices, choices I *must* make, is wholly upon me. (3) It follows that I must dwell in a *value-world of my own creation*, a world without the support of God or absolute values to guide me, a world wherein the notions of equity, justice, and mercy are simply magical fancies I use to comfort me. (4) So the historical forces I myself am caught up in—the Western tradition in which I see myself participating—clearly cannot be anything like a historical document of how Reason is working toward its absolute *telos*, as Hegel and the German Idealists believed; rather, history is simply the documentation of human failure in our hopeless quest for absolutes. (5) It likewise follows that any attempt to deliver my ultimate responsibility for choice over to some higher authority is an act of fundamental cowardice, of "inauthenticity," "bad-faith," or "superficiality." (6) Clearly, then, I must fully admit if I am to be totally honest with myself, the world in which I dwell can have no value, no meaning, no sense to it apart from what I give it myself. (7) This must be true for the others whom I clearly know to populate the world. They are *there* but their value is wholly opaque to me. It is I and I alone who must invest the other with value, meaning, and even human dignity.

The consequences of our interpretation of Pascal corresponds closely to Sartre's depiction of Angst and the Self in the face of "existentialist" freedom. And certainly, many traces of Heidegger's analysis of Dasein's everyday, inauthentic structures resemble this view. Moreover, as Sartre admits, his notion of Angst-filled freedom is based on Kierkegaard's seminal conception of Angst. So Kierkegaard at least by implication fits into our interpretation, especially when we consider his work *The Present Age*./7/

But our description lacks an important dimension at least from Tillich's standpoint. Tillich might agree: our description of Angst in the face of the Self's absolute freedom is something like his depiction of the Self's "ontological anxiety," experienced as the threat of non-being over the abyss of meaninglessness and death. Nonetheless, Tillich would argue that our description lacks the ontological and dialectical polarity of *destiny*. His is, after all, a *finite* freedom standing ontologically opposed to Sartre's *absolute* freedom. Surely, Tillich might ask, doesn't the finite self also *participate* in Being as well as stand *in opposition* to Being through the negativity and the Nothingness of human freedom? Moreover, Tillich might ask: Isn't it so that our description fails to recognize *courage* as the answer to the devastating ontological threat we describe here? The courage to affirm my Self even "in-spite-of" non-being's daunting yet fascinating challenge to human being is for Tillich the answer to Pascal's haunting vision. Tillich might conclude, against our interpretation of Pascal, that *it is sheer arrogance* to see ourselves as the source of values, for man's freedom is finite. To be an existent being *means* to exist in a world of pre-given values. The world is there: a historical, value laden realm of ontic concerns, a world *already there* in a cultural and historical setting. Hence, my world belongs to me precisely because I have a fully developed Self that *can* transcend every possible environment. Thus my Self participates in the world, and indeed in Being itself, in a delicate dialectical balance between freedom and destiny. To sacrifice my destiny, as would seem to be necessary from our extension of Pascal, means I must flee thoughtlessly into a chimera of absolute freedom. In the face of my totally finite being,

Sartre's absolute freedom is a patent contradiction. To be *absolutely* free requires infinite and absolute Being. Man is that being that intuits his radical finitude through the apodicticity of his self consciousness. We cannot be both finite and absolutely free.

Finally, Tillich would argue that in order to exist we *must* either affirm our Selves or else give in to the despair of non-being. The only way I can "carry-on" is through courage once the Angst of freedom is disclosed to me, to say "yes!" to Being "in-spite-of" non-being's threat to my ontic, spiritual, and moral affirmations of Self. The issue can only be resolved for me in faith—faith in the "God above God" who emerges when Angst has stripped the traditional notions of God from me.

For Tillich, then, Sartre has failed to grasp the essence of reflective Angst, because he ignores or forgets the ontological and dialectic of polarities of *both* freedom and destiny. In man this polarity exists as finite freedom in tension with finite destiny. To ignore finite destiny as does Sartre is to snap the thin thread that extends over the abysses of fate and death, emptiness and meaninglessness, and finally of guilt and condemnation.

Thus while Tillich and Sartre agree with Pascal that reflective Angst is the threat of non-being to the Self, they clearly disagree between themselves on the implications of the Self. Specifically: (1) For Sartre the Self is absolutely free; for Tillich it is free only up to the limits of its finitude. (2) For Sartre the Self is the source of all values; for Tillich the Self participates in a value-laden realm of being. (3) For Sartre the Self is isolated in a *de trop* realm of being. The Self is itself an object of consciousness, i.e., pure freedom, pure nothingness. For Tillich the Self participates as an ontological being suspended between freedom and destiny.

But despite their differences Sartre and Tillich agree on at least one point. Dasein, as an ontological being, is the basis for the Self: Dasein is the only being who can question Being. To prove this point let us compare Sartre's and Tillich's views on this matter directly. First, in *Being and Nothingness* Sartre says:

> . . . appearance of the self beyond the world—that is, beyond the totality of what is real—is an emergence of "human reality" [Heidegger's Dasein] in nothingness. It is in nothingness alone that being can be surpassed. At the same time it is from the point of view of beyond the

> world that being is organized into the world, which means on the one hand that human reality rises up as an emergence of being in non-being and on the other hand that the world is suspended in nothingness. Anguish (*angoisse*) is the discovery of this double, perpetual nihilation. It is in terms of this surpassing of the world that Dasein manages to realize the contingency of the world; that is to raise the question, "How does it happen that there is something rather than nothing?" (*BN*, p. 51).

Tillich also believes in the ontological as well as the logical priority of pre-reflective Dasein to the notion of Self. For example, in *Systematic Theology* he states:

> Man occupies a pre-eminent position in ontology, not as an outstanding object among other objects; but as that being who asks the ontological question and in whose self awareness the ontological answer can be found. . . . "Philosophers of life" and "Existentialists" have reminded us in our time of this truth on which ontology depends. Characteristic in this respect is Heidegger's method in *Sein und Zeit*. He calls Dasein [sic] ("being there") the place where the structure of being is manifest. But Dasein is given to man within himself. Man is able to answer the ontological question himself *because he experiences directly* and *immediately* the structure of being and its elements. (*ST*, I:169) (Italics added.)

This agreement between Sartre and Tillich suggests a clear joint dependence upon a more primordial pre-reflective level of man's being; a level both ontologically and logically prior to man as a Self. Therefore, the Angst experienced at this deeper level must likewise be the more primordial. Pre-reflective Angst is the ontological rock bottom condition that in the final analysis must undergird both Sartre's *angoisse* description and Tillich's discussion of "ontological anxiety."

Pre-Reflective Angst and the Abyss

We come then to the second type of Angst suggested above which must be distinguished from the reflective Angst of the Self's freedom. This second or deeper level is what we called "pre-reflective Angst in the face of the abyss." Here again the two polar characteristics of the daunting and the fascinating elements must be present. Once again we will call upon thinkers not previously cited to display the daunting and the fascinating dimensions of pre-reflective Angst, to reveal its full ontological power.

First, Angst's daunting pre-reflective power at the very edge of the abyss has been captured by Miguel de Unamuno, who in *The Tragic Sense of Life* cites the Italian poet Leopardi's last lines of the *Song of the Wild Cock*:

> A time will come when this Universe and Nature itself will be extinguished. And just as of the grandest kingdoms and empires of mankind and the marvelous things achieved therein, very famous in their own time, no vestige or memory remains to-day. So, in like manner, of the entire world and of the vicissitudes and calamities of all created things there will remain not a single trace, but a naked silence and a most profound stillness will fill the immensity of space. And so before ever it has been uttered or understood, this admirable and fearful secret of universal existence will be obliterated and lost./8/

But the daunting and indeed *haunting* power of Leopardi's entropic view of "universal existence" is counterbalanced by the fascination that existence holds for pre-reflective Dasein: if life and joy are to exist at all they must exist now—right here and now—in the spontaneous affirmation of "universal existence," the only Being we can know. The stillness that Leopardi sees which "will fill the immensity of space" mutely beckons Dasein to saturate the universe with meaning *now* before the certainty of entropy sets in. As though in direct response to Leopardi's dark vision, Nietzsche, that intrepid affirmer of life, states with equal eloquence on the side of fascination:

> All the beauty and sublimity which we have attributed to real or imaginary objects I claim as the property and creation of man. They are his most beautiful justification. Man the poet and the thinker! Man as God, as love, as power! With that royal generosity has he impoverished himself and made himself feel miserable in order to worship things. Up to now his greatest baseness has been that he admired and venerated things, forgetting that it was he who had created what he admired. Reject the humble expression "Everything is subjective." Say rather: "It is our work! Let us be proud of it."/9/

Nietzsche does not refer to the Self as the source of *mankind*'s creative role in the universe of meaning. Indeed it is *man as ontological being* who creates meaning and values in a kind of pre-reflective spontaneity. Man as Dasein stands at the abyss of death, meaninglessness, and condemnation—spontaneously creating bridges across the chasm. Only man is capable of saturating Being with meaning. If there is any crassness in him, therefore, it is either in his failure to take responsibility for the values he creates or in his forgetting that such values are created by him. Ontological man forfeits to the gods a universal order of value which then takes on an independent status from man. Dasein was originally the creator of the gods and the transcendental values for which they stand, according to this view.

The Self as depicted by Sartre and Tillich is wholly excluded from this picture. The focus of both Unamuno's and Nietzsche's vision points to pre-reflective Dasein, the Dasein of which Heidegger directly speaks, as well as the "dreaming spirit" of Kierkegaard. Thus Heidegger, who sees Dasein as that being who questions Being, recognizes that the Being-question itself depends on primordial Angst as the *existential epoché:* the *only* Archimedian point from whence Being *can be questioned.* Without the pre-reflective lucidity of primordial Angst, no such question is or can be possible. Certainly Leibnitz's question, "Why is there anything at all rather than nothing?" is similar to Job's question to Elohim discussed in Chapter I: it is decidedly not a mere request for information. Rather, it is an explicit recognition of the abyss between Being and non-being. It is precisely this recognition which gives rise to the question of Being.

Kierkegaard's morally innocent, ontologically ignorant "dreaming spirit" can leap into the abyss only *because* pre-reflective Angst ". . . is a qualification of dreaming spirit" (*CA*, p. 41). The dreaming spirit becomes a Self only after a qualitative leap, a leap made possible by pre-reflective primordial Angst.

For both Kierkegaard and Heidegger, therefore, primordial Angst is a pre-reflective rather than a reflective dimension of being human. Specifically Kierkegaard sees such Angst is a theological category arising from God's prohibition against eating from the tree of knowledge, a prohibition which ultimately brought Adam to Angst: "The prohibition induces in him anxiety [Angst], for the prohibition awakens in him freedom's possibility."/10/ Moreover, as we have seen, for Kierkegaard Angst is freedom's actuality as the possibility of possibility. Adam is not a Self who reflexively sees himself as being *affected* by Angst in any reflective way. Rather Adam is primordial man, a schematized metaphor for pre-reflective man, the same concept schematized by Heidegger as pure *Da-sein*. This, we submit, is the secret connection between Kierkegaard and Heidegger, a nexus revealed *only* in their respective treatments of primordial Angst. Heidegger says man becomes Dasein through primordial Angst. Likewise Kierkegaard asserts the "dreaming spirit" emerges into existence through primordial Angst. In both cases Angst is pre-reflective rather than reflective, and is related to Being rather than the Self, as Self is conceived by either Sartre and Tillich.

We are not arguing, however, that the notion of Self is totally absent in Kierkegaard and Heidegger. To be sure, the Self plays an important role in the writings of both thinkers. For Kierkegaard the entiresequel to *Begrebnet Angest*, namely, *The Sickness unto Death*, deals specifically with the problem of the Self and despair. For Heidegger the Self is clearly the authentic mode of Dasein's being, the answer to the "who?" of Dasein, and a question to which Heidegger devotes the whole of Part I, Chapter IV of *Sein und Zeit*.

What we *are* claiming is that primordial Angst, and the pre-reflective apprehension which describes it, must logically and ontologically precede any notion of Self in either Kierkegaard's or Heidegger's thought. This macro-hermeneutical interpretation is, we submit, one grounded firmly in our micro-hermeneutical analyses of both Chapters II and III above. Consequently our interpretation reveals the deep but subtle relationship between Kierkegaard's and Heidegger's revelations of primordial Angst, namely, the pre-reflective apprehension of the abyss between finite and infinite Being in Heidegger's case, and between "dreaming spirit" and the guilty existential man of Kierkegaard.

Consequences of the Foregoing Descriptions

We may now ask: What consequences can be drawn from this hermeneutical interpretation?

First, the reflective apprehension of the Self in Angst, as disclosed by Sartre and Tillich, is possible only because of the pre-reflective Dasein undergirding any notion of Self. Thus reflective Angst must be a constituted or second order phenomenon, a kind of "epiphenomenon" that gives ontological significance to an ethical or *praxis* oriented "existentialism." From this we may conclude pre-reflective Angst *is* the very self-same primordial Angst we seek: the primordial Angst revealed ontologically in the face of the abyss between finite being and infinite Being.

Second, if this is so then the notion of Angst in the face of the Self and freedom is, phenomenologically speaking, a vacuous concept. More specifically, our analysis has shown that reflective Angst cannot be separated logically or ontologically from pre-reflective Angst: the pre-reflective is the basis for the reflective dimension of primordial Angst. Thus the psychoanalytic schools which attempt to reduce Angst to a wholly immanent human phenomenon, one related to a spurious conceptual schema such as the ego-id-superego manifestations of Self, can never decide what is a symptom of Angst and what is Angst itself./11/ Further, postulating the existence of a subconscious mind, much less an unconscious-mind, must begin an infinite regress into progressively more abstract notions of consciousness as the basis of primordial Angst. Finally, an immanence doctrine of Angst presupposes there exists a causal relationship between the origin of Angst and its symptoms, an origin which in some sense can be predicted and controlled. This is patently unverifiable, a perspective grounded in scientism's positivistic presuppositions that cannot account in any adequate way for purely human phenomena such as love, concern, care, boredom, much less the ontological dimension of Angst itself.

The third consequence is that Angst's role in Western philosophical thought, while being *grounded* in phenomenological ontology, has ontic significance in Sartre's and Tillich's "existentialist" doctrines—specifically, their doctrines concerning finite choice, responsibility for such choices, and the universalization of such choices. As we have seen, existential choices are grounded in reflective Angst for these thinkers. The ontic consequence of primordial Angst turns out to be an ethical order of resolute Daseins, responsible for their actions to one another in a shared world of human meaning. All this is, of course, grounded in an ontological conception of pre-reflective Angst.

Finally, a further consequence of major significance emerges from our analysis which, as far as we know, has not been specifically seen before: primordial Angst is what makes possible an understanding and interpretation of human existence from within the purview of *aletheia*, the truth of Being. Heidegger observes that only in Angst can we become Dasein, and Dasein is that being which stands in the truth of Being (*SZ*, pp. 221–23). Thus we would argue that *Angst is the hermeneutic phenomenological counterpart to Husserl's "pure" phenomenological epoché*; it is the Archimedian point, so to speak, from which Dasein can discover the truth of Being. We believe further that his interpretation is confirmed in "What is Metaphysics?" and especially in the Postscript to that work where Heidegger insists that Angst reveals the Nothingness that is the condition of the possibility of the finite transcendence. *Primordial Angst is, therefore, the existential epoché of hermeneutic phenomenology.*

While this role cannot be extended to Kierkegaard, Sartre, or Tillich without some fancy hermeneutical footwork, both Sartre and Tillich base most of their own descriptions and interpretations of Angst on Heidegger's analysis. Hence, in a manner of speaking there may be some carry-over of the existential *epoché* notion to Sartre's depiction of the *en-soi* and Tillich's depiction of human being in the face of non-being's threat. We decline, however, to stretch the point beyond what we have said here for the sake of completeness.

Primordial Angst: A Broader Horizon

We come now to the questions toward which all our labors have been directed. First, what is the ultimate condition inherent within the Dasein that expresses Dasein's essential unity of understanding, disposition, speech, and fallenness? Second, how is this inherent condition revealed in primordial Angst?

Indeed, we must not shrink back from these haunting questions, even in the face of the profound awe and terror they may engender regarding the dark roots of human existence. But to inquire into these matters is to leave behind the guidelines of Heidegger's hermeneutic phenomenology and set sail now toward uncharted seas into the deepest levels of Dasein. Yet if we are to plumb the abysmal *bathos* of Angst's murky and fathomless depths, we must set sail without the familiar coastline to guide us.

Our analyses of Angst's many facets seem to rest upon one fundamental assumption that heretofore has remained unexamined: Dasein is a being for whom its finitude is regarded as a fundamental *privation* of some ontological wholeness. As Dasein I pre-reflectively apprehend in primordial Angst my incommensurability with infinite Being. This I regard as not only mere privation, but an axiological or value-laden privation, schematized as the Fall of Man from an original state of grace to the existential state of sin and dis-grace. Job and Prometheus, as we saw in our first chapter, are decidedly not interested in merely passively acquiescing their finitude and impotence without fundamental ontological protest. Indeed, it is their purpose to seize Being itself and to demand an account of it, and in this attempt to possess Being and make it their own: Job *demands* that Elohim speak; Prometheus knows the yoke of Ananke and in this knowledge becomes her true master.

Thus at the origins of the modern Western tradition, the very confluence of the Hebrew and Greek traditions, we see a basic element of Dasein's being emerging from the abyss of primordial Angst: Dasein is a being whose *primary* mode of being is ontological craving—the burning desire to assert its transcendence over beings despite its finitude, the burning desire to stand out into the infinite by transcending the world of objects, the burning desire to master Being itself. This, we submit, is a primal craving (*Urwille*) behind Job's demands upon Elohim. But more to the point, it is also the subtle power that motivates Dasein to pose the question: "Why is there something rather than nothing?" Like Job's demands, the question of Being is no mere request for information. Rather, the question's fundamental power and purpose is revealed within the question itself—the explicit demonstration that Dasein not only *can* ask the question, but that Dasein's "quest" for its answer *is*, in fact, the history of metaphysics in the Western tradition.

Dasein, therefore, is a "questing" as well as a "questioning" being. The tools for carrying out the quest are Dasein's ontological structures: understanding, pre-disposition, and speech. But the concept of fallenness or forfeiture, the fourth of Dasein's ontological structures, delimits the power of the quest. The question, which is the articulation (*Rede*) of the quest's power, goes well beyond Dasein's fallenness or finitude, extending into the ontological realm itself. Thus primordial Angst is the pre-reflective apprehension of the impossibility of any satisfactory answer to the question and, indeed, to the ultimate quest that stands behind it.

What makes possible the quest itself, the very condition of the possibility of Dasein, is the primordial craving, the impetus to human transcendence. Without such craving Dasein would be simply another being in a world of beings, the totality of which could be comprehended only by a God. Thus, Angst, as we have described it here, is a pre-reflective apprehension which penetrates to the darkest abyss lying between Dasein and the question of Being. Primordial Angst brings to light my craving, which in turn reveals to me not only my doomed quest for Being, but finite limitations preclude any satisfactory answer.

This concept of finitude craving (*Urwille*) is, of course, not one new to the Western tradition./12/ The term is Jacob Boehme's. But Arthur Schopenhauer also saw its power, and Nietzsche's conception of the will to power is clearly in the tradition we are suggesting here. It would appear, therefore, that the *Urwille* is a kind of pre-ontological hunger, a passion, a *craving* for ontological fullness. Recognizing the pre-mythopoeic, pre-ontological, and pre-reflective power of "metaphysical craving" establishes an essential element of primordial Angst, namely, its passionate and urgent nature. Thus we are not interested here in anything like a bloodless notion of will such as the dispassionate notion of "the problem of free will." Rather this pre-reflective and passionately non-rational craving for ontological fullness is precisely what Western philosophy has sought to cover-over, bury, and hide in the bone-yard of dubious passions. What reflective, contemplative, and rationalistic Western metaphysics fails to grasp is that if, as Plato suggests, philosophy begins in wonder, then the "quest" of the question of Being owes its origins not only to the pre-mythopoeic craving for understanding but to primordial Angst as the impetus for such understanding.

We need not look far in the Western tradition to ground this interpretation. In fact, we need not look beyond what has been shown in the micro-hermeneutical analyses of the present work. Behind primordial Angst's power lies its precondition: the pre-ontological craving for Kosmos in the face of the abyss. This is further borne out in our discussion of primordial Angst's four facets discussed in our first chapter.

Onto-theological Angst, for example, the Angst that stands behind mythopoeic thought, finds its sources in a passionate craving for ontological wholeness. Such Angst is the pre-reflective apprehension, therefore, of my frustrated craving to complete the synthesis of creatures and Creator. Hence, mythopoeic thought passionately provides the synthesis as the bridge across the abyss of onto-theological Angst.

In the Greek tradition of Ananke and strictural Angst, primordial Angst is interpreted as the restrictions placed on Greek Dasein's ontological craving to grasp the whole of Being as Kosmos. Ananke, as we saw in our first chapter, binds *more* than the intellect; it holds fast the passionate will and desires of Greek Dasein. Thus the schematizing concepts of Ananke as ontological necessity, binding, choking, strangling, etc., are mute hermeneutical testimony to the primordial craving in Dasein for ontological wholeness. Indeed the entire notion of "binding necessity" makes little sense unless there is something to bind. We suggest that it is the ontological craving of Greek Dasein, who through primordial Angst understands and interprets these restrictions as those of ontological necessity.

In pneumatical Angst, the facet of primordial Angst manifested in gnostic Dasein, we find enumerated the precise number of barriers, the 365 spheres of the evil Archons, standing between the *pneuma* and ontological wholeness: Dasein's original home beyond the cosmos. This deep desire, this open craving to transcend homelessness, can only be grounded in a pre-ontological *Urwille* to return home to the fullness of Being. Gnostic Dasein, therefore, saw the doom of finitude via pneumatical Angst: the clear pre-reflective apprehension of the literal abyss (Hell) into which the *pneuma* is cast—yearning, craving, passionately hungering for spiritual comfort and relief from the tyrannical rule of the evil Archons.

While such craving becomes less obvious in mystical Dasein's passional Angst, it is certainly no less intense an experience, as the accounts of Paul, Augustine, Eckhart, and Boehme indicate. But the passionate mysticism of St. John of the Cross, the *Ascent of Mount Carmel* and the *Dark Night of the Soul*, for example, are resounding metaphorical tributes to a primal craving to the abandonment which characterizes passional Angst. This is no patent contradiction, the craving to abandon will; for ontological craving is fundamentally different from ontic will. Thus the *dark night of the soul* as metaphor reflects a deep, passional Angst that pre-reflectively apprehends the abyss between the soul and the ultimate Godhead. The "dark night" metaphor displays that behind Eckhart's abandonment (*Abgeschiedenheit*) lies a primal craving to unite with infinite Being. Therefore ontological craving must be the full impetus behind and beneath mystical Dasein's quest to unite with the source of its being.

Finally, the craving as *Urwille* is specifically articulated by Jacob Boehme. As we have seen, in Boehme's thought the dark forces of the *Ungrund* (the abyss), the *Urwille* (conceived by Boehme as God's primordial will), and the *Urmensch* (primal man, both before and after the Fall) are all brought together into a doctrine of ontological craving or desire in Being itself. The *Urwille* for Boehme is the ontological craving *of* the Godhead for meontic self-knowledge.

Even in Kierkegaard's analysis of Angst as sympathetic antipathy and antipathetic sympathy we discover Angst is secretly dependent on the notion of an ontological craving for what one is apprehensive about. This interpretation is made explicit in a 1842 *Journal* entry where Kierkegaard observes: ". . . anxiety [Angst] is an alien power that *grips the individual*, and yet one cannot tear himself free from it *and does not want to*, for one fears, *but what he fears he desires*."/13/ (Italics added.)

Sartre makes absolutely no secret concerning ontological craving. His entire analysis of the "existentialist" structures of the *en-soi* turn around the *en-soi*'s frustrated craving to become one with the *pour-soi*, Being itself. Regarding Tillich the entire basis of ontological courage is a craving for ". . . the God who appears when God has disappeared in the anxiety of doubt" (*CB*, p. 190). Thus, the "existentialist" literature on Angst, whether it be theistic or atheistic, gives full ascent to ontological craving for the wholeness of Being. But the notion of such ontological craving becomes obscured by Sartre and Tillich's joint insistence upon an interlarded reflective Self between Angst and the abyss.

But what about Heidegger? Have we not simply eliminated him from this analysis? Not so. The specter of *Being and Time* hovers like the silent call of conscience behind every word written here. It is therefore time to make explicit how Dasein can crave the ontological fullness and how Angst reveals the limits of such craving, from Heidegger's perspective.

Dasein is the only being who can even pose the question of Being, much less quest to answer it. This is because Dasein is a thrown-projection, a temporal being whose authenticity is anticipatory resoluteness toward its final *telos*, death. In Angst's existential *epoché*, inauthentic being-in-the-world is grasped by me as the Nothingness of the world's totality-of-possible concerns (*Bewandtnisganzheit*). This grasping makes possible my freedom to transcend inauthenticity and become a full partner with Being as authentic Dasein.

Authentic Dasein is, we suggest, the closest man can come to uniting with ultimate Being. For *in* such authenticity, man renounces for the first time his ontological craving for an impossible equality with Being. Man finally accepts his finitude as resoluteness toward death. *But only through primordial Angst can Dasein become authentic.*

Here the spirit of gnosticism drifts through Heidegger's thought—urging it, compelling it to a richness previously not possible in the Western tradition; for as Kierkegaard saw Hegel's grand final synthesis of thought and Being, Absolute Reason's full teleological fruition must exclude anything like a Dasein as an individualized and fully finite being-in-the-world. Man's finitude prevents the synthesis; man's existential condition is necessarily finite. The gnostic notions of (1) the call, (2) Dasein's alienness in the everyday world, (3) the metaphysical homesickness discovered in Angst's ontic "uncanniness," (4) the vision of man as a thrown being, and (5) even the notion of the Angst engendered by man's separation from Being: each of these themes moves in and through Heidegger's *Being and Time* and "What is Metaphysics?" with the same covert power that the Bible provides for *Begrebnet Angest*.

Primordial Angst contains *within itself* the full ontological meaning of human finitude; for the notion of a "pre-reflective apprehension of the abyss between finite and infinite Being" describes the grounds for the apprehension: an ontological craving for the fullness of Being announced mutely in Dasein's Angst-filled recognition that such fullness can never come to pass for man as a finite being.

If we have adequately worked through the daunting dimension of primordial Angst's basic conditions, namely, the vision of Dasein's essential craving for absolute Being delimited by primordial Angst, then we must also display at a macro-level Angst's fascinating dimension wherein craving is itself transcended and becomes meditative thinking—a final horizon wherein Being and Dasein's craving for being come together.

Angst and Gelassenheit

There is, of course, no "solution" to the "problem" of primordial Angst any more than there can be a "solution" to the "problem" of human finitude. Indeed there *is* no problem of Angst; there is only the daunting as well as the alluring intensity with which Angst first overtakes us in ontic "uncanniness."

In the preceding section we discussed the daunting power of primordial Angst. In the present section our task is to show how Angst can be transformed into a fascinating, creative, and dynamic mode of existing, a way of authenticating being-in-the-world that renounces craving for ontological wholeness as an intentional act. This intentional act we call "renunciation." Through renunciation Being reveals itself to Dasein, via the difficult task of "meditative thinking," as Heidegger understands that term.

So as we began, so shall we return to Heidegger's philosophy for one final sweep around the hermeneutic circle. We are afforded an opportunity in Heidegger's later thought to bring our discussion of Angst's role to a close by showing how Angst points ahead to new and exciting philosophical and metaphilosophical horizons, horizons which transcend the representational thinking of the Western philosophical tradition. "Pointing ahead" means there can be no closure to the question of primordial Angst and its relationship to Being. Rather, there can only be a revitalized exploration of even vaster horizons wherein Being's meaning may be questioned. This having been stated, let us examine as our final horizon Heidegger's notion of *Gelassenheit* in relation to primordial Angst.

In 1959, two and one-half decades after "What is Metaphysics?" Heidegger published *Gelassenheit*, translated by John M. Anderson and E. Hans Freund as *Discourse on Thinking.*/14/ The work consists of two parts: a Memorial Address in honor of the German composer Conradin Kreutzer, and a dialogue entitled "Conversations on a Country Path," between a scholar, a scientist, and a teacher.

The term *Gelassenheit* is another German expression that is extremely difficult to render into English. According to Hoffmeister's *Worterbuch der Philosophischen Begriffe* [*Dictionary of Philosophical Concepts*],/15/ *Gelassenheit* comes from the Middle High German word *Gelazen* which means: "to settle down" or "to take up one's abode." Specifically Hoffmeister defines "*Gelassenheit*" as follows:

> . . . an expression from mysticism and Pietism meaning the serenity [*Ruhe*] in God acquired by the full renunciation of earthly affairs. Today it generally means: firmly grounded in authentic Being (*eigenden Wesen*)./16/

Thus while Anderson and Freund translate *Gelassenheit* as "releasement," we submit that "renunciation" (in the sense of a serene renouncing of the world of ontic concerns) is closer to what Heidegger means by *Gelassenheit*. This means that Heidegger's *Gelassenheit* is, in a very important sense, a continuation of *Being and Time*: Heidegger works out of the question of Being in *Gelassenheit* from within the standpoint of authentic or resolute Dasein, the Dasein which, because of primordial Angst's existential *epoché*, can relinquish everyday ontic concerns and can now approach the Being question from within Being's own horizons.

The argument of *Gelassenheit* appears deceptively simple: the history of the Western intellectual tradition, says Heidegger, has been conveyed through a restless, calculative, and discursive mode of thinking that seeks to force beings into the canons of reason. But there is another *kind* of thinking that does not seek to manipulate or "re-present" beings. This second mode, called "meditative thinking" (*besinnliches Denken*), may be characterized as "hearkening to the voice of Being." Hearkening to Being's voice allows Being to reveal itself to Dasein through Dasein's participation in meditative thinking. Moreover, meditative thinking is inherent in the nature of man, says Heidegger./17/

But what is it that must be renounced in meditative thinking? In the "Memorial Address," Heidegger says the thoughtless glorification of the calculative thinking that permeates our contemporary technological civilization must be renounced so that we may see technology in its true light. "I call the comportment which enables us to keep open to the meaning hidden in technology," says Heidegger, "openness to the mystery."/18/ Behind calculative thinking, the technological and discursive thinking that focuses solely upon beings, stands the mystery technology seeks to overcome. In the "Conversation on a Country Path,"

Heidegger goes a step further: what is renounced is representational (calculative) thinking, a kind of willing to represent beings in terms of human reason. Representational thinking forces the things-themselves to reveal themselves to us in human terms, namely, through our categories of reason. We never see beings on their own terms in representational thinking.

To renounce representational thinking is to renounce its source, my will to grasp, transform, and understand beings on my own terms. This does not mean calculative thinking must be wholly abandoned, however; rather, what must be relinquished is the all-embracing hold calculative or technological thinking has over us in contemporary thought. To get from calculative to meditative thinking, says Heidegger, we must will *not to will*, so that *Gelassenheit* ". . . awakens when our nature is let-in so as to have dealings with that which is not a willing."/19/ Only on the basis of renunciation, as a willing not to will, can meditative thinking take place in Dasein at a spontaneous or pre-reflective level.

Some of the important characteristics of meditative thinking discussed by Heidegger are: Meditative thinking (1) is not representational, that is, it does not reconstruct a world of objects in our minds; (2) begins with an awareness of a horizon wherein beings reveal themselves; (3) is open to Being's content within the horizon, and does not force categorical concepts upon what shows itself to us within that horizon; (4) is not an act of self-will to grasp in gestalt the wholeness of Being; rather, meditative thinking renounces self-will completely in a higher act of will, one characterized by an openness to Being; (5) is not passive but rather requires the active and constant courage to maintain its pristine clarity; (6) is a higher type of activity than the craving of representational thought; (7) is the manner in which Dasein approaches being and Being approaches Dasein, namely, Being reveals its horizons to Dasein; (8) is dwelling in the realm of Being, a kind of serving Being which authentically lights up Being's horizons; (9) is an openness to Being which discloses what is hidden or veiled in Being; and finally, (10) is a resolve for truth that makes possible the disclosure of Being on Being's own terms rather than on human terms.

Heidegger insists that meditative thinking is spontaneous pre-reflective thought. This means the craving of representational thought can never lead Dasein to the wholeness and fullness of

Being. In other words, meditative thought is the true direction of the *Urwille.*

We believe the meaning of representational thinking, which "... computes ever new, ever more promising and at the same time more economical possibilities ...," which "... races from one prospect to the next ...," and which "never stops, never collects itself,"/20/ is grounded in the ontological *craving* discussed in our preceding section. Heidegger gives this ontological craving the more philosophically acceptable word, "will."

Meditative thinking, however, gets to Dasein's roots, Dasein's true grounding in Being itself. Only in the primordial act of renouncing *craving* for Being can Dasein get beyond representational thinking. For at last we see the point: *representational thought, the primary thought mode of traditional western metaphysics, is the abyss itself.* Consequently, the deeper we penetrate into the secrets of beings, the further Being is separated from us. The harder we *will* Being to show itself, the more quickly Being flees from us. Clearly the *Urwille* must reverse its ground in thinking of Being if Dasein is to make any progress at all in its quest for Being.

At the very dawn of mythopoeic Dasein, at the very primitive stirrings of primordial Angst, before man could even think or "reason," there emerged, "... a kind of 'gut-level' apprehension of an absolute groundlessness of Being expressed as an abyss; a feeling that there may be no meaning to existence in the lived-world of experience. ..."/21/ We argued in Chapter I that ontotheological Angst arose from this primordial Angst when the transcendental totality of beings was interpreted as a Supreme Being "... which then becomes schematized as a pantheon of deities or a monotheistic God."/22/ When this schematization first took place was the very beginning of representational thought, for Dasein does not worship, question, or defy the transcendental totality of beings—Being itself, rather, Dasein worships, questions, or defies the schematized icons that stand for such a lofty ontological conception. These are called "gods."

Thus the abyss widens and deepens with each successive layer of dogma and ritual surrounding philosophical metaphysics and theological speculation. Little wonder the gnostics sought their original home beyond this hell called Earth. For them the transcendental totality of Being has slipped away in the 365 spheres between Dasein and the high God.

Eckhart saw the only way out of the labyrinth between beings and Being. This was total "detachment" (*Abgeschiedenheit*). Detachment is related to Eckhart's notion of the inner man who ". . . has in his essence the inner strength to completely detach himself from the world. . . . Such detachment takes the form of a self-renouncing abandonment of worldly concerns attained in a destruction of self-will."/23/

Yet Heidegger's conception of *Gelassenheit* goes beyond Eckhart's detachment. Says Heidegger:

Scientist:	The transition from willing into . . . [*Gelassenheit*] is what seems difficult to me.
Teacher:	And all the more, since the nature of . . . [*Gelassenheit*] is still hidden.
Scholar:	Especially so because even if . . . [*Gelassenheit*] be thought of as within the domain of will, as is the case with old masters of thought such as Meister Eckhart.
Teacher:	From whom, all the same, much can be learned.
Scholar:	Certainly; but what we have called . . . [*Gelassenheit*] evidently does not mean casting off sinful selfishness and letting self-will go in favor of divine will.
Teacher:	No, not that./24/

In Heidegger's account renunciation goes beyond any human conception of divine will. It is a renunciation of both representational thought and the abyss *of reason* that makes representational thinking possible. Renunciation is a direct or pre-reflective intuition into the ground and abode of Being, an insight which undercuts all dichotomies of finite-infinite, temporal-eternal, or creature-creator. Heidegger's *Gelassenheit* abandons not only these dichotomies and the representational concepts to which they point, but far more importantly, *Gelassenheit* means to abandon the small hyphen between these dichotomies: those uncomplicated symbols used in ordinary language which contain within their purview the final abyss of primordial Angst.

To renounce representational thinking in no way implies we have "overcome" primordial Angst. As we have attempted to show throughout this chapter, Angst's existential *epoché* as the condition of authentic existence means that *the ultimate authenticity of*

Gelassenheit can take place only from within primordial Angst. Angst must again be affirmed as primordial pre-disposition, a primary attunement to Being which makes possible meditative thinking. Indeed Angst *is* Dasein, authentic Dasein as meditative thinker. To become authentic we must seek not to overcome primordial Angst, but to *live* authentically and meditatively *within* Angst's existential *epoché*.

It seems clear that since meditative thinking requires a higher will to renounce the ontic will of representational thinking, the notion of *Urwille* cannot be abandoned. In this renunciation, primordial Angst becomes an authentic openness to Being, and an authentic pre-reflective understanding of what Being gives to Dasein. From all of this there emerges the vision of pure *Da-sein*, spontaneously interacting with Being in what Heidegger calls the "region of all regions." Spontaneous, pre-reflective apprehension is apprehension not in the sense of the first definition provided above, namely, "fear as to what might happen; dread." Rather, in delivery over into the region of Being, apprehension becomes "the ability to understand, understanding." In *Gelassenheit* the *Urwille* has finally found rest; for at last it has transcended the realm of beings and dwells now in the house of Being.

A New Beginning—The Final Sweep of the Hermeneutic Circle

In this essay we do not mean to suggest that philosophy and the representational thinking which is its primary articulation (*Rede*) should be overthrown by a radical and meditative revolution in thought. As Heidegger has observed, representational thinking has its lasting place in the being of man. Neither are we advocating that the *Urwille*, primordial Angst, and Dasein can be transcended. Moreover, we are not advocating a "philosophy of feeling," a doctrine of the irrational, or any other attempt to undercut the discursive philosophical enterprise from its deep historical roots.

To be sure, we appreciate the place of representational thinking and the technological advances it has brought us in virtually every field of human endeavor, from the dawn of Greek hylozoism to beyond the contributions of Heisenberg. We are not so arrogant as to deny technology's enormous accomplishments or their intrinsic philosophical, spiritual, and even ethical value. The voice of representational thinking rings clearly in the *logos* of Being.

We are saying, however, that in the clear night of primordial Angst we discover representational thinking and the technology, science, and the underlying philosophy it serves to be insufficient to satisfy the *Urwille* to the Being question. Nowhere in canons of representational thought can there be an answer to the question of Leibnitz: "Why is there something, rather than nothing?" The canons of reason do not admit of a satisfactory answer.

And so we have attempted to show the major role Angst can play in Western philosophical thought: namely, to provide access to meditative thinking. We can come to meditative thinking and *Gelassenheit* only after working through the history of Western philosophy from mythopoeic to meditative thought. Angst itself, as the pre-reflective apprehension of the abyss of representational thought, shows the way to *Gelassenheit*. At last Angst becomes the medium of *Gelassenheit* itself.

Our work here is only a call to accept into the pristine towers of academic philosophy the visions and intuitions revealed by primordial Angst. Rather than advocating a doctrine of the irrational or the overthrow of logic and reason, we seek only to enrich and supplement thought itself by recognizing that primordial Angst can light up a realm of human being. To light up human being (Dasein) is simply not possible if we cling to the canons of reason as the only mode of philosophical inquiry. As we have tried to show, such clinging is a form of ontic will, that when clutched to the breast in a desperate embrace, can only smother the truth of Being—the ontological truth that rushes in upon us when in *Gelassenheit*,

WE LET GO!

Surely he who confuses *techne* with *sophia* has little claim to *philo-sophia*; for wisdom is far more than technique or technology. The quest for wisdom may require many paths through the forest of thought, and our small effort here has been simply to point out another way: the way from mythopoeic to meditative thinking through the opening in the woodlands which separates them, the opening of hermeneutic phenomenology—a path on the way to meditative thinking.

We do not claim to have transcended primordial Angst, the *Urwille*, or Dasein in this essay. We suspect that beyond these primordial manifestations of human being, none but a god dare go. We have no desire to philosophize with a hammer or with an axe, but only to point to the authentic freedom primordial Angst discloses as the existential *epoché*. We seek only the serenity of *Gelassenheit*: to dwell in the voice and house of Being, a dwelling made possible only within primordial Angst itself. And thus we come around the arc of the hermeneutic circle for the final sweep by allowing our primary guide, Martin Heidegger, the occasion of our conclusion.

> The clear courage of primordial Angst vouches for the most mysterious of all possibiities: the experience of Being. For close to primordial Angst—as the terror of the abyss—abides awe. Such awe, illuminates and covers-over each dwelling place of mankind, within which he comfortably abides in the abiding./25/

NOTES

/1/ *The Oxford English Dictionary*, rev. ed., s.v. "apprehension."

/2/ Supra.

/3/ Supra.

/4/ *ST*, I:107.

/5/ *ST*, I:170.

/6/ Blaise Pascal, *Pensees*, translated by W. F. Trotter as *Thoughts* (the Harvard Classics, Vol. 48; New York: P. F. Collier and Son Company, 1910), p. 73.

/7/ Søren Kierkegaard, *The Present Age*, translated by Alexander Dru and Walter Lowrie (New York: Oxford University Press, 1940).

/8/ Miguel de Unamuno, *The Tragic Sense of Life*, translated by J. E. Crawford Flitch (New York: Dover Publications, Inc., 1954), p. 124.

/9/ Robert G. Olson, *An Introduction to Existentialism* (New York: Dover Publications, Inc., 1962), p. 40. Note: Olson does not cite the source from Nietzsche.

/10/ *CA*, p. 44.

/11/ *CB*, p. 65.

/12/ We are thinking here primarily, although certainly not exclusively, of Arthur Schopenhauer's notion of the will as endless striving which in Copleston's words is ". . . a blind urge or impulse which knows no cessation, it cannot find satisfaction or reach a state of tranquility." Cf. Frederick Copleston, S. J., *A History of Philosophy*, Vol. 7, Part II (Garden City, New York: Image Books, 1963), p. 38. But this tradition clearly goes back as far as Boehme.

/13/ *JP*, Vol. 1, 94 (III A 233), n.d., 1842.

/14/ Martin Heidegger, *Gelassenheit* (Pfullingen: Guenter NESKE Verlag, 1959), translated by John M. Anderson and E. Hans Freund as *Discourse on Thinking* (New York: Harper & Row, 1966).

/15/ Johannes Hoffmeister, *Woerterbuch der Philosophischen Begriffe*, Zweite Auflage (Hamburg: Felix Meiner, 1955).

/16/ Ibid., p. 253. Note: the present author is wholly responsible for this translation.

/17/ Heidegger, *Discourse on Thinking*, p. 58.

/18/ Ibid., p. 55.

/19/ Ibid., p. 61.

/20/ Ibid., p. 46.

/21/ Supra.

/22/ Supra.

/23/ Supra.

/24/ Heidegger, *Discourse on Thinking*, pp. 61–62.

/25/ Heidegger, *Nachwort*, p. 103/ *G*, p. 307.

BIBILIOGRAPHY

Aristotle. "Metaphysics." *The Works of Aristotle*. 2d ed. Translated by W. D. Ross. Oxford: Clarendon, 1940.

Astruc, Alexander, and Contat, Michel. *Sartre by Himself*. Translated by Richard Seaver. New York: Orison Books, 1978.

Augustine. *The Confessions of St. Augustine*. Translated by E. B. Dussey. New York: Airmont, 1969.

Baader, Franz X. von. "Vorlesungen über religiouse Philosophie." *Sammtliche Werke* 1–16. Liepzig, 1850–60.

Barnes, Hazel E. Translator's Introduction to *Being and Nothingness* by Jean Paul Sartre. New York: Pocket Books, 1956.

Beardsley, Monroe. *Thinking Straight: Principles of Reasoning for Readers and Writers*. 4th ed. Englewood Cliffs, NJ: Prentice-Hall, 1975.

Biemel, Walter. *Martin Heidegger: An Illustrated Study*. Translated by J. L. Mehta. New York: Harcourt Brace Jovanovich, 1976.

Billeskov, Jansen, F. J. *Studier i Søren Kierkegaards Litteraere Kunst*. [*Studies in Soren Kierkegaard's Literary Art*.] Copenhagen: 1958.

Blackham, H. J. *Six Existentialist Thinkers*. New York and Evanston: Harper Torch Books, 1959.

Brock, Werner, ed. *Existence and Being*. Gateway Edition. Chicago: Henry Regnery, 1949.

Caputo, John D. *The Mystical Elements of Heidegger's Thought*. Oberlin: Ohio University Press, 1978.

Clark, James M. *Meister Eckhart: An Introduction to the Study of His Works and an Anthology of His Sermons*. London: Thomas Nelson and Sons, 1957.

Copleston, Frederick. *A History of Philosophy*, 9 vols. Garden City, NY: Image Books, 1962–72.

Cornford, F. M. *From Religion to Philosophy: A Study in the Origins of Western Speculation*. New York: Harper & Row, 1957.

_________ . *Plato's Cosmology: The Timaeus of Plato*. London: Routledge, 1948.

Dilthey, William. *Gesammelte Schriften*, 3. Stuttgart: Teuber Verlag, 1958–61.

Eckhart, Meister Johannes. *Meister Eckhart: Selected Treaties and Sermons*. Translated by J. M. Clark and J. V. Skinner. London: Farber & Farber, 1958.

Frankfort, Henri; Frankfort, Mrs. H. A.; Wilson, John A.; and Jacobson, Thorkild. *Before Philosophy*. Baltimore: Penguin Books, 1964.

Fromm, Erich. *Escape from Freedom*. New York: Avon Books, 1965.

Gadamer, Hans-Georg. *Truth and Method*. New York: The Seabury Press, 1975.

_________ . *Wahrheit und Methode*. 2d ed. Tübingen: Paul Sieback Verlag, 1965.

Grant, Robert M. *Gnosticism: A Source Book of Heretical Writings from the Early Christian Period*. New York: Harper & Brothers, 1961.

Gray, J. Glenn. *The Warriors: Reflections of Men in Battle*. New York: Harcourt Brace, 1959.

Grene, Marjorie. *Sartre*. New York: New Viewpoints, 1973.

Harkness, Gloria. *Mysticism: Its Meaning and Message*. New York: Abington Press, 1973.

Hegel, George W. F. *Lectures on the Philosophy of Religion*, 3 vols. Translated by E. S. Haldane and Frances H. Simon. London: Routledge & Kegan Paul, 1963.

_________ . *Sämtliche Werke, Jubiläumsausgabe*, 20 vols. Edited by Hermann Glockner. Stuttgart: Frommann, 1927–39.

Heidegger, Martin. *Being and Time*. Translated by John Macquarrie and Edward Robinson. New York: Harper & Row, 1962.

_________ . "Brief über den Humanismus," *Platons Lehre von der Warheit*. Bern: Franke, 1947.

_________ . *Discourse on Thinking*. Translated by John M. Anderson and E. Hans Freund. New York: Harper Torchbooks, 1966.

———. Introduction to *Heidegger: Through Phenomenology to Thought* by William J. Richardson. The Hague: Martinus Nijhoff, 1963.

———. *Kant und das Problem der Metaphysik*. 2d ed. Frankfurt: Klostermann, 1951.

———. "Letter on Humanism." Translated by Frank A. Capuzzi, in David Krell, ed., *Martin Heidegger: Basic Writings*. New York: Harper & Row, 1977.

———. "Nachwort zu: 'Was ist Metaphysik?'" *Gesamtausgabe*, Band 9. Frankfort: Klostermann, 1976.

———. *On Time and Being*. Translated by Joan Stambaugh. New York: Harper & Row, 1972.

———. *Sein und Zeit: Erste Hälfte*. 6th ed. Tübingen: Neomarius Verlag, 1949.

———. *The Essence of Reasons*. Translated by Terrance Malick. Evanston: Northwestern University Press, 1969.

———. *Vom Wesen des Grundes*. 4th ed. Frankfurt: Klostermann, 1955.

———. *Was ist Meaphysik?* 5th ed. Frankfurt: Klostermann, 1949.

———. "Wegmarken." *Gesamtausgabe*, Band 9. Frankfurt: Klostermann, 1975.

Hillman, James. "On the Necessity of Abnormal Psychology." *Eranos*, vol. 42 (pp. 91–135). Leiden: Brill, 1947.

Hoffmeister, Johannes. *Wörterbuch der Philosophischen Begriffe*. Zweite Auflange. Hamburg: Felix Meiner Verlag, 1955.

Hong, Howard V., and Hong, Edna H., eds. and translators. *Søren Kierkegaard's Journals and Papers*, 7 vols. Bloomington and London: Indiana University Press, 1967–78.

———. Translator's foreward, in *Kierkegaard's Thought* by Gregor Malantschuk. Princeton: Princeton University Press, 1971.

Hopper, David. *Tillich: A Theological Portrait*. Philadelphia and New York: J. P. Lippincott, 1968.

Husserl, Edmund. *Ideas: General Introduction to Pure Phenomenology*. Translated by W. R. Boyce Gibson. New York: Collier Books, 1962.

Inge, William R. *Mysticism in Religion*. London: Hutchinson's Universal Library, n.d.

Jonas, Hans. *The Gnostic Religion*. 2d ed. Boston: Beacon Press, 1963.

Jones, Rufus M. *Studies in Mystical Religion*. London: Macmillan, 1909.

Katsaros, Thomas, and Kaplan, Nathaniel. *The Western Mystical Tradition: An Intellectual History of Western Civilization*. New Haven: College & University Press, 1969.

Kaufmann, Walter. *Discovering the Mind*, vol. 2. New York: McGraw-Hill, 1980.

Kegley, Charles W., and Bretall, Robert W. *The Theology of Paul Tillich*. New York: Macmillan, 1952.

Kelley, C. F. *Meister Eckhart on Divine Knowledge*. New Haven and London: Yale University Press, 1977.

Kierkegaard, Søren. *Concluding Unscientific Postscript*. Translated by David F. Swenson and Walter Lowrie. Princeton: Princeton University Press, 1941.

_________ . *Either/Or*, 2 vols. Translated by Walter Lowrie. Princeton: Princeton University Press, 1944, 1959.

_________ . *Fear and Trembling* and *The Sickness Unto Death*. Translated by Walter Lowrie. Princeton: Princeton University Press, 1954.

_________ . *Repetition*. Translated by Walter Lowrie. Princeton: Princeton University Press, 1941.

_________ . *Søren Kierkegaards Papirer*. Edited by P. A. Heiburg and Victor Kurr. 2d expanded edition. Edited by Niels Thulstrup. 25 vols. Copenhagen: Gyldendal, 1968–78.

_________ . *The Concept of Anxiety*. Translated by Reidar Thomte. Princeton: Princeton University Press, 1980.

_________ . *The Concept of Dread*. 2d ed. Translated by Walter Lowrie with an Introduction and Notes. Princeton: Princeton University Press, 1957.

_________ . *The Present Age*. Translated by Alexander Dru and Walter Lowrie. New York: Oxford University Press, 1940.

Kramer, Samual N. "Man and His Gods." *Ancient Near Eastern Texts*, Supplement. Princeton: Princeton University Press, 1955.

Krell, David, ed. *Martin Heidegger: Basic Writings*. New York: Harper & Row, 1977.

Lambert, W. G. *Babylonian Wisdom Literature*. Oxford: Oxford University Press, 1960.

Loew, Cornelius. *Myth, Sacred History and Philosophy: The Pre-Christian Religious Heritage of the West*. New York: Harcourt, Brace & World, 1967.

Mackey, Louis, *Kierkegaard: A Kind of Poet*. Philadelphia: University of Pennsylvania Press, 1971.

Macquarrie, John. *Existentialism*. Harmondsworth, Middlesex: Penguin Books, 1977.

Mahan, Wayne W. *Tillich's System*. San Antonio, TX: Trinity University Press, 1974.

Martin, Bernard. *The Existentialist Theology of Paul Tillich*. New York: Bookman Associates, 1963.

May, Rollo. *Paulus: Reminiscences of a Friendship*. New York: Harper & Row, 1973.

_________ . *The Meaning of Anxiety*. Revised Edition. New York: W. W. Norton, 1977.

_________ . "The Origins and Significance of the Existentialist Movement in Psychology." In *Existence: A New Dimension in Psychiatry and Psychology*, edited by Rollo May, Ernest Angel, and Henri F. Ellenberger. New York: Simon & Schuster, 1967.

May, Rollo; Angel, Ernest; and Ellenberger, Henri F., eds. *Existence: A New Dimension in Psychiatry and Psychology*. New York: Simon & Schuster, 1958.

McBride, William Leon. "Man, Freedom, and *Praxis*." In *Existential Philosophers: Kierkegaard to Merleau-Ponty*, edited by George A. Schrader, pp. 261–329. New York: McGraw-Hill, 1967.

McCarthy, Vincent A. *The Phenomenology of Moods in Kierkegaard*. The Hague and Boston: Martinus Nijhoff, 1978.

Mehta, J. L. *Martin Heidegger: The Way and the Vision*. Honolulu: University of Hawaii Press, 1976.

_________ . *The Philosophy of Martin Heidegger*. Varansi: Banaras Hindu University Press, 1967.

Nordentoft, Kresten. *Kierkegaard's Psychology*. Translated by Bruce H. Kirmmse. Pittsburg: Duquesne University Press, 1978.

Olson, Robert G. *An Introduction to Existentialism*. New York: Dover Publications, 1962.

Otto, Rudolf. *The Idea of the Holy: An Inquiry into the Non-Rational Factor in the Idea of the Divine and Its Relation to the Rational*. Translated by John W. Harvey. London: Oxford University Press, 1958.

Owens, R. B. "The Knees of the Gods," *The Origins of European Thought* (pp. 303–9). Cambridge: Cambridge University Press, 1954.

Pascal, Blaise. *Thoughts*. Translated by W. F. Trotter. *The Harvard Classics*, vol. 48. New York: P. F. Collier & Son, 1910.

Pfeiffer, Franz P. *Meister Eckhart*. Translated by C. de B. Evans. London, 1924.

Plato. "Timaeus." *The Works of Plato*. Translated by B. Jowett. New York: The Dial Press, 1936.

Pritchard, J. P., ed. *Ancient New Eastern Texts Relating to the Old Testament*. 2d ed. Princeton: Princeton University Press, 1955.

Przywara, Erich. *An Augustine Synthesis*. Gloucester: Peter Smith, 1970.

Rad, Gerhard von. *Genesis: A Commentary*. Translated by John H. Marks. Philadelphia: The Westminster Press, 1961.

———. *Wisdom in Israel*. Nashville and New York: Abington Press, 1972.

Richardson, William J. *Heidegger: Through Phenomenology to Thought*. The Hague: Martinus Nijhoff, 1963.

Robinson, James M., ed. *The Nag Hammadi Library*. Translated by members of the Coptic Gnostic Library Projects. San Francisco: Harper & Row, 1977.

Rolt, C. E. "Mystical Theology." *On the Divine Names and Mystical Theology*. New York, 1920.

Rosenkranz, Karl. *Psychologie oder die Wissenschaft vom Subjecktiven Geist*. Konigsberg, 1937.

Sallis, John, ed. *Heidegger and the Path of Thinking*. Pittsburgh: Duquesne University Press, 1970.

Sartre, Jean Paul. *Being and Nothingness*. Translated by Hazel E. Barnes. New York: Pocket Books, 1956.

———. *Existentialism and Human Emotions*. New York: Philosophical Library, 1957.

———. "Existentialism is a Humanism." In Walter Kaufmann, *Existentialism from Doestoevski to Sartre*, pp. 278–311. New York: Meridian Books, 1956.

———. *Le Être et le néant: Essai d'ontologic phenomenologique*. Paris: Gallimard, 1940.

_________ . *Search for a Method.* Translated in part by Hazel Barnes. New York: Knopf, 1963.

_________ . *Sketch for a Theory of the Emotions.* Translated by Philip Mairet. London: Methuen, 1962.

_________ . *The Transcendance of the Ego: An Existentialist Theory of Consciousness.* Translated and annotated by Forest Williams and Robert Kirkpatrick. New York: The Noonday Press, 1957.

_________ . *The Words.* Translated by Bernard Frechtman. New York: Vintage Books, 1981.

Schreckenberg, Heinz. "Ananke: Untersuchungen zur Geschichte des Wortgebrauchs." *Zetemata*, Heft 36. Münich: Beck, 1964.

Scott, R. B. Y. *The Way of Wisdom in the Old Testament.* New York: Macmillan, 1971.

Seidel, George J. *Martin Heidegger and the Pre-Socratics: An Introduction to His Thought.* Lincoln: The University of Nebraska Press, 1964.

Sibbern, F. C. *Bemaerkninger og Undersogelser, fornemmelig betraeffende Hegels Philosophi, betraget i Forthold til vor Tid.* [*Observations and Investigations Particularly on Hegel's Philosophy, Seen in Relation to Our Time.*] Copenhagen, 1838.

_________ . *Om Erkjendelse og Granskning.* [*On Knowledge and Research.*] Copenhagen, 1822.

Speigelberg, Herbert. *The Phenomenological Movement*, 2 vols. The Hague: Martinus Nijhoff, 1965.

Steiner, George. *Martin Heidegger.* New York: Penguin Books, 1978.

Stoudt, John J. *Jacob Boehme: His Life and His Thought.* New York: The Seabury Press, 1968.

Taubes, Susan A. "The Gnostic Foundations of Heidegger's Nihilism." *The Journal of Religion* 33–34 (1954): 155–72.

Tillich, Hannah. *From Time to Time.* New York: Stein & Day, 1973.

Tillich, Paul. *Das System der Wissenschaften nach Gegenständen und Methoden.* Gottingen: Vandenhoeck & Ruprecht, 1923.

_________ . "Existential Philosophy." *Journal of the History of Ideas* 5 (1944): 44–70.

———. *My Search for Absolutes*. New York: Simon & Schuster, 1967.
———. *Mysticism and Guilt Consciousness in Schelling's Philosophical Development*. Translated by Victor Nuovo. Lewisburg, PA: Bucknell University Press, 1974.
———. *On the Boundary: An Autobiographical Sketch*. New York: Charles Scribner's Sons, 1966.
———. *Systematic Theology*. 3 vols. in 1. Chicago: University of Chicago Press, 1951–63.
———. *The Construction of the History of Religion in Schelling's Positive Philosophy*. Translated by Victor Nuovo. Lewisberg, PA: Bucknell University Press, 1974.
———. *The Courage To Be*. New Haven and London: Yale University Press, 1952.
Unamuno, Miguel de. *The Tragic Sense of Life*. Translated by J. E. Crawford Flitch. New York: Dover Publications, 1954.
Underhill, Evelyn. *Mysticism: A Study in the Nature and Development of Man's Spiritual Consciousness*. New York: Sutton, 1961.
Versenyi, Laszlo. *Heidegger, Being and Truth*. New Haven and London: Yale University Press, 1965.
Wahl, Jean. *Jules Lequier*. Paris: Editions des trois Collines, 1948.
Wyschogrod, Michael. *Kierkegaard and Heidegger: The Ontology of Existence*. New York: Humanities Press, 1969.